The Science of Creationism

By Lawrence A. Wilcox

Published Amazon's Kindle

Bible scripture quotations (marked in gray boxes) and citations were taken from the King James Version of the Bible.

Amazon Publishing

ISBN: 9798375650463 (Paperback)

Table of Contents

The Science of Creationism

Introduction

Christian Creationism is straight from the Holy Bible. For this book, the story of the creation is quoted from The King James Version of The Holy Bible. The last two chapters are a brief discussion of Intelligent Design as suggested by today's religious leaders. This book does not consider other religious narratives of the beginning of earth and time.
The purpose of science is verifying hypothesis and expanding the knowledge to apply to similar studies. The creation of earth cannot be replicated to verify the source, so theories have to be evaluated with probabilities. Questions over the Biblical quotations and analysis are given at the end of each chapter to allow the student to evaluate the Biblical narrative and in some cases the religious interpretation. The probability of the Christian version of creation is not a popularity contest between creationism and evolution or validated by evangelists' rhetoric. The correctness of the Christian Creationism is independent and devoid of the probability of evolution being correct.

The probability for support or contradiction of the Biblical Creation is evaluated at the end of each chapter by questions at the end of the chapter. Probability is determined by estimating the creditability of statements in lieu of actual verification. The scale of measure will be to assign a number of 1 to 5 for each statement. A total contradiction to a statement will be 1, an implausible statement will be a 2, a possible and maybe even practical statement will be a 3, a plausible and likely statement will be a 4, and a confirmed or verifiable statement will be a 5. At the end of each chapter, answer each question or statement with the level of probability of its accuracy.

Sum the probabilities for each chapter and add the sums of each chapter to obtain the total for the book. Or it may be desirable to separate chapters 1 through 14 on the Biblical Creation from chapters 15 and 16 on Intelligent Design. For the book total divide by the total number of evaluations, 115 to get the probability of creation according to the scriptures in

The Holy Bible and Intelligent Design. A better analysis is to evaluate the Biblical Creation separate from the Intelligent Design. A reason to separate the two sources is the Bible has existed many centuries and Intelligent Design is a very recent strategy. The Biblical Creation has 100 questions to average and the Intelligent Design has 15 questions to average. Evaluating the probability of the creation story being plausible makes public education assess the creditability of the Bible which violates the separation of church and state. To not evaluate the probabilities of occurrences fails to complete the investigation of the course and fails to fully teach students the subject. Not evaluating evolution creates the opportunity for deceit by religions to deny it's a science. But the evaluation of evolution is left for science courses and not related to creationism.

Determining the probability that the creation story occurred as stated in the Bible or as stated by Intelligent Design, may create political and religious displeasure but the academics virtues exceed the risk of displeasing Christian leaders.

This is an objective factual investigation, not a faith in a deity investigation. The facts are evaluated exempt of spiritual truths. The difference in past and current levels of knowledge should not obscure objective evaluation. It is now known the earth is roughly spherical and orbits the sun that spirals in the Milky Way galaxy as the galaxy travels outward in an expanding universe. Writers of the text that were later canonized into the Bible could only perceive the earth as flat. This does make the creation story wrong when the story is evaluated as a factual event. The evaluation must be devoid of sympathy for the writers and not evaluating it as a spiritual truth, or a level of knowledge of the writers, scribes and editors of the Bible. Consideration for the writers' lack of knowledge does not change the probability of their correctness. Facts are used from NASA photographs and measurements made by scientists, technicians and university studies. Citations are not given as facts are available from internet searches.

The quotes from the Bible are boxed with gray background.

Chapter 1: Creating Heaven and Earth

Book of Genesis Chapter 1

1:1 In the beginning God created the heaven and the earth.
1:2 And the earth was without form, and void, and darkness was upon the face of the deep. And the Spirit of God moved upon the face of the waters.
1:3 And God said, Let there be light: and there was light.
1:4 And God saw the light, that it was good: and God divided the light from the darkness.
1:5 And God called the light Day, and the darkness he called Night. And the evening and the morning were the first day.
1:6 And God said, Let there be a firmament in the midst of the waters, and let it divide the waters from the waters.
1:7 And God made the firmament, and divided the waters which were under the firmament from the waters which were above the firmament: and it was so.

"…God created the heaven and the earth." This was written long before it was discovered that the earth was a sphere. The 'earth is flat' was the general knowledge when Genesis 1 was written. The belief that the earth was flat persisted through the many times it was scribed on new papyrus, parchment, leather, or linen, and through many translations. Heaven was considered to be the domain of Gods and existed above the earth. Establishing the earth and heaven gave a common starting platform for explaining the existence of all things. Note that Bible does not state God created the universe first and then our sun. Stating heaven and earth were separated also created domains for God and human beings. Having heaven above earth established the hierarchy for God being in heaven above the earth and above humans that are to be created later in the story. This is similar to mythologies brought back from Egypt, and Assyria with the exception that the Hebrew faith was dedicated to being monotheistic instead of polytheistic.

As the story starts, the earth was without form, and void, and darkness was upon the face of the deep. With no understanding of physics, this was a method early writers used to convey a sense of chaos, disorder and likely a fear of the unknown. With a minimal understanding of the world, nighttime and darkness were frightening and not subject to the small amount of order and limited security added by vision. The beginning with only the surface of water and everything dark resembles the Egyptian religious mythology of the beginning of the earth. The difference is that Egyptian mythology is polytheistic and the first God, Re (or Re-Atum aka Ra-Atum) rose from the deep and dark chaotic water to establish order and begin creating the world. The Egyptian mythology developed along the Nile, which had a great influence on the story development. The Nile's annual flood brings life to the land so it was felt that the Nile or water was the source of life. The influence was strong enough to carry over to the use in the creation story brought back with the Jews on leaving Egypt.

"And the Spirit of God moved upon the waters" conveys several messages with subtle implications from obscure wording. The use of "Spirit" allows the reader to infer several different interpretations of the actions taking place. Spirit may imply an aberration floating and moving effortlessly and silently with no physical restrictions. Having spiritual qualities unbounded by physical laws elevates God's spirit above human frailty. The aberration could be viewed as God Himself or a spirit dispatched by His whim. Either way the statement portrays God as above the physical laws governing humans. Spirit could also represent the presence of God without the actual physical presence of God. This meaning of spirit could be much the same that is in use today. An example of usage today would be if a great philanthropist were to die after creating a holiday tradition of feeding the needy and fellow associates were to continue the celebration and traditional holiday feast. It could be said his spirit could be felt at the dinner. Again, this connotation implies writers of Genesis believed God to be of superior powers.

Moving on the face of the waters has several subtleties. With the absence of light or order, God or His Spirit moving through the newly formed heaven-earth conglobation of chaos implies an initiation of order and structure. The water being represented as a face, as opposed to a surface, gives an animate characteristic to the water. This again shows the strength of God over water as He was in the water's face.

"And God said, Let there be light: and there was light." This statement has two purposes. The first is showing that God created with just a command, demonstrating the ultimate power. Recall that at the time these verses were written and copied there was no scientific understanding of light, so creating light was a phenomenal act. Second is adding intelligence and order to chaos by adding light. Separating light from darkness again demonstrates control and increases the order in the new existence. Evaluating the creation of light at this point by saying, "God saw the light, that it was good" creates a huge dilemma. Either the Bible is the inerrant word of God or it is ancient human religious writings. If this statement is the inerrant word of God, it presents at least one of two contradictions of being from the infinite power and wisdom.

The statement, "God saw the light and that it was good", has two inconsistencies with being the inerrant word of God. The possibilities of God's judgment of light are that God was amazed by light and had to evaluate it for us, or that God was showing pride in His work and had to reveal this to us. Was God amazed by light? This is not probable. The creation story does not state if God has always existed for an infinite length of time or if God instantly appeared with all His power. If God existed forever, prior to His creating light did He always have all His power or did it grow over time. Did it require infinite time to acquire infinite power? The creation story does not explain God's existence or the creation of God. God's amazement with light is contradictory with infinite power and wisdom. If God has always existed, it is not possible that He existed for an infinite time in total darkness. Maybe time did not exist before God started the creation. The starting of time brings up the question of what existed before time, but this is too complex and deep to ponder when there is a Bible to evaluate. The basic fact is that God being amazed by light is not consistent with extreme power and wisdom to create the universe. Then being amazed by light would seem contradictory to an all-powerful God. If God felt compelled to evaluate light and stated this only to indicate to the readers that light was good then this is insulting to the readers. They are using light to read, using the safe assumption that the first writings were not in Braille[1]. Judging light for the sake of the readers seems a bit unintelligent for a God.

If God was proud of His work not just amazed by it, His stating His pride creates another problem. God stating, "that it was good" as pride would set precedence. A slight divergence to analyze the actions of God will help establish the creditability of the story. Pride is considered one of the seven deadly sins. The seven deadly sins or any sins, laws, or rules have not been established in the Bible for the story. Sins not being established in the story does not change their impropriety. Behavior is either ethical or unethical based on its virtue not whether laws, rules or religious tenets describe the action, so God's actions are still crucial to human development even if the story has not included the definition of sins. Later in the story of the creation, God's action of resting on the seventh day was adapted as a religious law so rigid that violation was a capital offense. God's actions are important even before the conveying of laws, rules and sins. Also, God's action of resting on the seventh day will establish that God sets rules by example. God's action may not be a sin because of the fact that He also decides right and wrong and judges who has violated His laws. God may exempt Himself or He has self-immunity. God's self-immunity does not eliminate His example setting policy unless He dictates policy counter to His actions. Or God leads by example except where He dictates laws contrary to His example. God did not tell us to ignore His actions in this story. Therefore, God's stating, "that it was good" was not pride.

The statement, "God saw the light and that it was good" is not consistent with an all-powerful God dictating His word for the Bible from either case analyzed. Eliminating the possibility that this is the inerrant word of God exposes the fact that the Bible is the writing of human beings. The Bible being written by humans is consistent with this statement and the technical knowledge of several thousand years ago. Light was not understood and would have seemed mystical. God judging light would not implicate a conflict in understanding since God was to be all-powerful and understanding light was not within the writers' capability.

Naming day and night and the separation, which created the first day, is a practical start for a story based on the knowledge level of several thousand years ago. There had to be a first day if there were an instant creation. If the sun grew from condensing gases and began combustion as internal pressures grew over several billion years while the earth simultaneously grew from collecting materials captured in the sun's orbit over several billion years, then the first day would be too nebulous to define. So, the beginning of the universe would be a much more complex scenario to describe, and thus deferred to other science courses.

"And God said, Let there be a firmament in the midst of the waters, and let it divide the waters from the waters" can only be understood if it is known what a firmament is. If it was known in Ancient Hebrew, somehow the meaning did not translate well. It was translated from the Hebrew 'ragiya' (pronounced rä·kē'·ah), which was used in Hebrew to mean a mountain or extremely large hill extending up from the plain. There is a second meaning of a dome or solid hemisphere which could be interpreted as the sky. The second meaning seems more likely to be the intended use. If it were a plain or solid like a mountain rising out of the water, this would coincide with what the writer has called earth. Defining a 'sky' would seem more likely and would create the mechanism the writers believed for the sun and moon to enter at the base and skim along the firmament in their travel each day. The firmament would also allow a suspension system for the stars. The added benefit of having this dome covering earth is that it provides a place for heaven, extending above the firmament.

Writers of the Tanakh (aka the Torah) felt compelled to initiate the stories from the beginning of everything in the known world. The tribal knowledge during this period was heavily influenced by Egyptian and Mesopotamian mythology. The Egyptian and many religions of the Sumerian area, i.e. Sumerian, Akkadian, Assyrian, and Babylonian, were polytheistic and had to be converted to monotheistic to align with Jewish religion. For the Tanakh, the creation had to result from one God's creation instead of each part being created by a separate god as in the other religions. But the overview of the creation is the same: God had to create heaven, earth and sky from the dark chaotic waters. The Biblical story extends a bit further to add God raising the firmament out of the waters. The Egyptian and Sumerian religions also had their Gods raising heavens or mountains out of the waters. In this way, the stories are quite similar.

"And God made the firmament, and divided the waters which were under the firmament from the waters which were above the firmament: and it was so." Does this change the meaning of firmament? Let us delay the evaluation of this statement until after further reading of Genesis.

Book of Genesis Chapter 1

1:8 And God called the firmament Heaven. And the evening and the morning were the second day.
1:9 And God said, Let the waters under the heaven be gathered together unto one place, and let the dry land appear: and it was so.
1:10 And God called the dry land Earth, and the gathering together of the waters called the Seas: and God saw that it was good.

"And God called the firmament Heaven. And the evening and the morning were the second day." This is pretty light day's labor for God, but we should not be judgmental. So, God called the firmament heaven. The fact is that light and dark determined the second day. This statement does further define the firmament as heaven, so the question about the waters above firmament or heaven still has no better refinement.

"And God said, Let the waters under the heaven be gathered together unto one place, and let the dry land appear: and it was so. And God called the dry land Earth, and the gathering together of the waters called the Seas: and God saw that it was good." The story began with God creating heaven and earth. Now God called the firmament that He raised out of the waters heaven and then He gathered the waters for seas and the dry land for earth. There is either a conflict in actions or the writers' clarification appears to create a duplication of tasks. The first line says God created heaven and earth, and statements 9 and 10 say He gathered earth as dry lands under the firmament, which He raised later out of the dark chaotic waters. Let's allow that God created earth and during the later processes of raising the firmament and setting light and dark in motion to create days, that earth required a bit of refinement, which was described as gathering it together. At this point we have earth and seas and the waters above the firmament for which God has not related the purpose. Just a pure speculation for fun, we may consider that the writers placed water above the firmament as a thought of heaven as an island paradise. However, island paradise is a more modern concept and not very likely to have been the intent of the writers.

Questions over Chapter 1:

1. God created the earth and heaven (sky above earth) before light
    a. because the writers were not knowledgeable of the universe and the order of stellar creations.
    b. because the writers saw their surroundings and felt it required a higher power than themselves to create it.
    c. because the writers adapted religions, traditional believes, or mythologies from other cultures into their creation story.
    d. could be any or all the above.
2. Heaven and Earth were dark and chaotic with no light
    a. to set a stage for God's bringing order.
    b. because darkness was frightening.
    c. because light had not been invented yet.
    d. all the above
3. The use of water covering everything symbolized that
    a. land had not been created yet so God's presence indicates super powers.
    b. it was known that water brought life as in the annual Nile flooding.
    c. land was still submersed and this set the conditions for the beginning of life.

    d. water was considered evil and brought death by floods.

4. The spirit of God could imply all but
    a. an aberration floating and moving effortlessly and silently with no physical restrictions.
    b. a spirit dispatched by God's whim.
    c. God's happy and joyous disposition.
    d. the presence of God without God's actual presence.
5. God created light and separated it from dark
    a. to create day and night.
    b. because the writers did not understand that ambient light for daytime was a result of the sun even if it were cloudy or the sun was unseen.
    c. as a show of strength.
    d. all the above.
6. As the writers perceived the creation, before creating light,
    a. God remained in the dark for billions of years.
    b. God had just recently created Himself so it was a very short period.
    c. God was so powerful He did not need light.
    d. all the above.
    e. none of the above.
7. God created the firmament
    a. which created the sky in a dome over earth and called it heaven.
    b. to separate earth from water.
    c. because the writers were unsure of how to separate waters from earth.
    d. all the above.
8. God gathered the waters for seas and the land for earth. This statement
    a. is different from the original opening statement that God created heaven and earth.
    b. is a deeper explanation of how God created Heaven and earth.
    c. shows the writers took a break and forgot the original statement.

d. is a restating of the first lines that gives the appearance of copying other mythologies that had gods separating waters and dry land.

Probability of statements in Chapter 1:

Instructions: Assign a number to the plausibility of each statement.

(1) For a total contradiction to a statement.
(2) For an implausible statement.
(3) For a possible and maybe even practical statement.
(4) For a plausible statement.
(5) For a confirmed or verifiable statement.

1. What is the probability that the writers of the Bible understood God, light, and physical laws and wrote of factual events not mythology or superstition? (1 – 5)?
2. The writers of the Bible wrote that God could create light without the sun. This showed their understanding of the universe, because they knew days could be light when overcast and start to lighten before sunrise. What is the probability that the earth could be illuminated without sunlight 3000 to 2000 years ago? (1 – 5)?
3. What is the probability that God's judgment of light as good was consistent with a super power capable of creating life? (1 – 5)?
4. What is the probability that God's judgment of light shows the Bible to be the inerrant word of God because it clearly defines His power and status above judgment? (1 – 5)?
5. What is the probability that the waters above the firmament had some relationship to the heavens and were not the imagination of the writers? (1 – 5)?
6. What is the probability that the writers of Genesis understood the structure of the universe and earth's place? (1 – 5)?
7. What is the probability that the Bible is the inerrant word of God, or God dictated the Bible in a manner that could be understood by unknowledgeable writers of the time? (1 – 5)?

8. What is the probability that God separating the waters and gathering the dry lands for earth is a deeper explanation of how God created heaven and earth? (1 – 5)?

Chapter 2: Adding Plant Life to Earth

> ### Book of Genesis Chapter 1
> **1:11** And God said, Let the earth bring forth grass, the herb yielding seed, and the fruit tree yielding fruit after his kind, whose seed is in itself, upon the earth: and it was so.
> **1:12** And the earth brought forth grass, and herb yielding seed after his kind, and the tree yielding fruit, whose seed was in itself, after his kind: and God saw that it was good.
> **1:13** And the evening and the morning were the third day.
> **1:14** And God said, Let there be lights in the firmament of the heaven to divide the day from the night, and let them be for signs, and for seasons, and for days, and years:
> **1:15** And let them be for lights in the firmament of the heaven to give light upon the earth: and it was so.
> **1:16** And God made two great lights, the greater light to rule the day, and the lesser light to rule the night: he made the stars also.
> **1:17** And God set them in the firmament of the heaven to give light upon the earth,
> **1:18** And to rule over the day and over the night, and to divide the light from the darkness: and God saw that it was good.
> **1:19** And the evening and the morning were the fourth day.

"And God said, Let the earth bring forth grass, the herb yielding seed, and the fruit tree yielding fruit after his kind, whose seed is in itself, upon the earth: and it was so." This would have been an excellent time in the story to also mention the banana. The banana is not unique in its situations but demonstrates a dilemma exceptionally well. For those not familiar with the scientific problem posed by the banana, a brief synopsis follows. The banana has no seeds and propagates through root expansion. It is possible at the very early stages of development the banana may have produced seeds, which now shows as five barely perceivable dark spots around the center of the fruit, which may be remnants of seeds. So, the banana as known today only propagates through root reproduction. The banana exists on every continent in slightly different varieties. The plants did not cross the oceans by root propagation, nor did animals or humans have the capability to transplant bananas from one continent to another. This leaves only two possible explanations. The reasonable and logical solution is that the continents were adjoined or connected at one time allowing the banana to spread over the whole massive single continent. Then, through plate tectonics, the continents separated into the familiar configuration we know today. Since the continental plates are still moving and this movement is measurable, this appears to be the most probable answer. The bananas must have propagated and spread across the joined continents after they evolved to a seedless fruit. There is no other explanation for the consistency that on all continents, that at lease one banana plant did not loose the ability to propagate through seeds. Since all bananas are alike in this aspect, it is most probable that they all lost this capability at the same stage in evolution and began evolving from this common start after the continents divided.

The second possible solution to the banana quandary is that God created a slightly different variety on every continent as a prank. Bananas have changed greatly in the past century due to selective breeding by Dole, Chiquita, Del-Monte, etc. to obtain fruit to meet commercial requirements. Prior to man's involvement with bananas, did God change the plants to be slightly different on each continent? This is not a plausible explanation. It would have been plausible if the story of creation had told this to be the case. Adding the story of the banana to the Bible would have certainly shown God's craftiness and sense of humor and would have been beyond the knowledge of the writers. The absence of the explanation of the banana phenomenon shows the limitation of the explanation of the variations of plant life around the world. The limited knowledge in the story argues against the Creation story as the inerrant Word of God and endorses the case for the Bible being written by humans. The limited knowledge in the story supports the idea that the creation story was constructed from stories, traditions, and collections from nearby lands and shows the supernatural influence or guidance to be a small probability.

Christian evangelists now tout the banana as a perfect creation by their God. Christian evangelists hype the 'perfect creation theory' attempting to diminish the effect of the contradiction created by the banana's existence and to manipulate the banana's existence into proof of existence of their God. They claim that God created the banana as the ultimate in 'user friendly' fruits. They list all the good attributes of the banana and claim only God could create such a well 'designed' fruit. The Christian evangelists publicize attributes of the banana from selective propagation by many produce growers and suppliers and ignore plants that produced small fruit or varieties not selected for commercial growing. Other trees left fossil imprints on every continent with the same dilemma that they propagated only through root expansion. Fresh water fish pose the same dilemma. Similar species have evolved slightly different on different continents but yet could not have crossed salt-water oceans. These were not mentioned in the bible, but we have not come to animals yet.

"And the earth brought forth grass, and herb yielding seed after his kind, and the tree yielding fruit, whose seed was in itself, after his kind: and God saw that it was good. And the evening and the morning were the third day." God covered the earth with vegetation and completed the third day of the creation. The focus of the creation story appears to be to convey God's power and wisdom through the creations. The ability to create life must have been unimaginable power for the writers and editors and scribes of the Torah or Pentateuch (first five books of the present Old Testament). The description of the plant life could have been added to illustrate the grandeur of the world. Granted that writing and scribing at the time of writing the Torah were more difficult than today's use of automated word processors or even typewriters, but we are discussing what is purported to be 'God's Word'. If the Bible is God's word, then omissions are just as revealing as what is included such as no explanation of life enhancing creations to the world. A possible explanation of God's pronouncement that light was good, is possibly He is proud of the creation of light. If God is proud of light, why was He modest about the creating of fruit and nut trees such as figs, dates, almonds, etc.? These trees and fruit baring plants that providing some pleasures to life may have demonstrated a more palatable environment in which humans were to be created would seem to be a natural point of pride and showing that He created some luxury beyond simple existence. Stating, "it was good" shows a higher probability of humans writing the Bible than God. Some description of the staples to sustain life may have added a sense of comfort level to the new world. Wheat, barley and flax were not mentioned as available as minimum substances for life, although they are grasses and may have been understood as such. Food and maintaining life were major concerns during the time of writing the Bible and a plausible expectation would be that staples other than 'grass' would have been mentioned. From a scientific perspective, the omission of descriptions of the plants and trees restricts the evaluation of the creation and certainly adds to the probability that humans wrote the Bible and it is not the inerrant word of God. After the writing of Genesis and a couple of millennia of cross-pollination and selective cultivation have produced greatly improved field crops, fruit trees, herbs, and even flowers and decorative plants. The agricultures, vineyards, orchards, and gardens of today reflect thousands of years of development. The scientific question is why did God not create these plants for people thousands of years ago?

"And God said, Let there be lights in the firmament of the heaven to divide the day from the night, and let them be for signs, and for seasons, and for days, and years: And let them be for lights in the firmament of the heaven to give light upon the earth: and it was so. And God made two great lights, the greater light to rule the day, and the lesser light to rule the night: he made the stars also. And God set them in the firmament of the heaven to give light upon the earth, And to rule over the day and over the night, and to divide the light from the darkness: and God saw that it was good. And the evening and the morning were the fourth day." These statements reveal some of the writers' and editors' lack of understanding of the solar system. The writers, editors and scribes did not know stars could be over 13,000,000,000 light years away. During the time the Bible was written, edited, and scribed on different media like papyrus and parchment, the skies were cleaner and clearer than today. Many more stars and galaxies would have been visible to the unassisted eye. They still would not have been able to detect distance galaxies and certainly not 13 billion light years, but the point is that the night sky must have been fascinating. The building of

Stonehenge was started about 3000BCE, and predates the final[2] writing of the Torah scripts that became the Bible. Stonehenge has alignment with astrological events, which indicates cognizance of solar and lunar motions. Stonehenge exhibits the capability to coordinate time to solar and lunar movements so it is not a stretch to image that writers of the Torah were also capable of understanding the relationship of seasons and days, and that seasons were annual. Again, the writers and editors refer to the firmament as the sky and supporting stars, sun and moon and they do not comment on the statement about the water above the firmament. Here the writers and editors have forgotten what was declared eleven statements earlier. God had already created light and separated light and dark to make day and night and God does it again with the creation and motion of the sun, moon and stars. This second account of the creation of day and night gives hints that the writers were copying mythology (religious beliefs) from Egypt and Mesopotamia. It appears the writers and scribes had to combine differing mythologies and were not careful to eliminate overlaps and contradictions. The mythologies brought back from these areas were polytheistic and considered the sun, moon and stars as Gods. Condensing the Jewish creation to a single God required the sun, moon and stars be reduced from Gods to objects created by God and kept God more powerful by giving Him domain over them. A reiteration with a different connotation was included for day, night, light, dark and the solar system.

"And the evening and the morning were the fourth day." Note that there were three days before the creation of the solar system, sun and moon, etc. God created the sun a day after creating all plants. Plants need sunlight to survive, and if He had delayed this longer, then the plants would have died and the work of the third day would have been lost and required redoing. This story reflects the understanding of the writers, editors and scribes. Very important is the fact that the writers thought of a day as the same length of time as we do today. This was demonstrated by not considering that the plants could be in peril from lack of light. Today interpretations of the creation are false that are used to equate the Biblical story to actual time of development by stating, a day to God may have been thousands or millions of years. The writers intended a day to equal a day as indicated by the lack of concern for creating plants a day before creating the sun. A more logical approach to the story of creation would have been to create the sun, moon and solar system, then earth and then start populating it with vegetation. Again, this reflects the level of knowledge used in writing the creation story and clearly shows what they intended a day to be in time.

Questions over Chapter 2:

1. God covered the earth with all varieties of plants
    a. to complete the third day.
    b. for man and animals to eat for survival when they were created.
    c. because the writers knew plants existed and had to include them.
    d. all the above.
2. God created plants
    a. to sustain life.
    b. to beautify the world.
    c. to establish an environment for humans.
    d. the story does not say why.
3. Plants have evolved through natural and human selection
    a. because God wanted life to start without better varieties.
    b. because the writers of the Bible knew nothing of the future.
    c. is an untrue statement from evolutionists.
    d. is a lie told by modern farmers to get more government subsidies.
    e. because God expected man to improve upon what he had been given?
4. On the fourth day God creates the sun, the moon and the stars
    a. to fill the firmament.
    b. because darkness was frightening.
    c. to set time references, hours, days months, etc.
    d. to create day and night.

e. as a means of communicating (giving signs) to man?

5. The third day God created minimal plants for sustaining life
    a. because this was a full day and God could not perfect plants.
    b. God did not want to make life too easy.
    c. because the writers of Genesis were not cognizant of future improvements.
    d. all the above.
6. Did writers of the "Old Testament" or Bible see the same sky as we do?
    a. yes, the sky has not changed.
    b. no, the air was much clearer.
    c. no, the middle east is in a different time zone.
    d. none of the above.
7. God did not explain that the earth was a sphere and orbited the sun because
    a. He wanted people to fear solar and lunar eclipses.
    b. He wanted people to fear meteorites and comets.
    c. other religions worshipped the sun and moon as Gods.
    d. the writers and editors knew nothing about the solar system.
    e. God wanted man to learn things for himself?
8. The story repeats creating light, day and night
    a. because the writers had no knowledge of physics.
    b. because the writers did not concern themselves with repetitions or logical flow of the story since it was written as religion not science.
    c. because there were many writers and scribes and they did not combine their story to support each other.
    d. it is likely the writers were copying other religions and tribal knowledge to document the creation for religious purposes.
    e. all the above.

9. The earth took over 4,500,000,000 years to develop.
    a. So a day in the creation was 643 million years.
    b. The day was the same length of time and scientists are wrong.
    c. The 6th day creation is one of the Spiritual Truths.
    d. The writers of the Bible had no idea how long it took for the earth to develop.

Probability of statements in Chapter 2:

Instructions: Assign a number to the plausibility of each statement.
(1) For a total contradiction to a statement.
(2) For an implausible statement.
(3) For a possible and maybe even practical statement.
(4) For a plausible statement.
(5) For a confirmed or verifiable statement.

1. Writers knew only of fruits that exist in the local world, in North Africa, Southeast Europe or the Middle East. Including all fruits of the world would have demonstrated knowledge exceeding their capability. What is the probability that this indicates God did assist in writing Genesis? (1 – 5)?
2. God made perfect plant life and human's genetic alterations (cross-pollination, selective reseeding, etc.) have deteriorated them. What is the probability that this is a true statement? (1 – 5)?
3. God's evaluation of the creation of all plants was that it was good. We now know the struggle that the human race has endured and the amount of genetic engineering that was required to develop crops to feed much of the earth but still better crops are required to feed all. What is the probability that God's evaluation was correct? (1 – 5)?
4. The story said God created the stars on the fourth day. What is the probability that God can place stars 14 billion light years away from earth? (1 – 5)?

5. The stories of creating heaven and earth, creating light and dark, separating light and dark, and creating the sun, moon and stars are repeated. What is the probability that this is because God wanted to emphasize these points in the creation? (1 – 5)?
6. A day in the creation was not long enough to endanger plants from absence of sunlight. The earth took more than 4.5 billion years to develop. What is the probability that writers intended a day to equal 643 million years or 4500 million divided by 7 to make the Bible agree with actual time? (1 – 5)?
7. Christian evangelists use the existence of the banana as proof of their God's creativity and beneficence and ignore the selective propagation by growers. What is the probability that that existence of the banana is proof of God's creativity and not selective propagation? (1 – 5)?

Chapter 3: Populating the Earth with Animals and Humans

> **Book of Genesis Chapter 1**
> **1:20** And God said, Let the waters bring forth abundantly the moving creature that hath life, and fowl that may fly above the earth in the open firmament of heaven.
> **1:21** And God created great whales, and every living creature that moveth, which the waters brought forth abundantly, after their kind, and every winged fowl after his kind: and God saw that it was good.
> **1:22** And God blessed them, saying, Be fruitful, and multiply, and fill the waters in the seas, and let fowl multiply in the earth.
> **1:23** And the evening and the morning were the fifth day.
> **1:24** And God said, Let the earth bring forth the living creature after his kind, cattle, and creeping thing, and beast of the earth after his kind: and it was so.
> **1:25** And God made the beast of the earth after his kind, and cattle after their kind, and every thing that creepeth upon the earth after his kind: and God saw that it was good.
> **1:26** And God said, Let us make man in our image, after our likeness: and let them have dominion over the fish of the sea, and over the fowl of the air, and over the cattle, and over all the earth, and over every creeping thing that creepeth upon the earth.
> **1:27** So God created man in his own image, in the image of God created he him, male and female created he them.
> **1:28** And God blessed them, and God said unto them, Be fruitful, and multiply, and replenish the earth, and subdue it: and have dominion over the fish of the sea, and over the fowl of the air, and over every living thing that moveth upon the earth.
> **1:29** And God said, Behold, I have given you every herb bearing seed, which is upon the face of all the earth, and every tree, in the which is the fruit of a tree yielding seed, to you it shall be for meat.
> **1:30** And to every beast of the earth, and to every fowl of the air, and to every thing that creepeth upon the earth, wherein there is life, I have given every green herb for meat: and it was so.
> **1:31** And God saw every thing that he had made, and, behold, it was very good. And the evening and the morning were the sixth day.

"And God said, Let the waters bring forth abundantly the moving creature that hath life, and fowl that may fly above the earth in the open firmament of heaven. And God created great whales, and every living creature that moveth, which the waters brought forth abundantly, after their kind, and every winged fowl after his kind: and God saw that it was good. And God blessed them, saying, Be fruitful, and multiply, and fill the waters in the seas, and let fowl multiply in the earth. And the evening and the morning were the fifth day." God populated the world with fish and birds. These statements give the meaning of firmament as everything above the earth such as the air for birds to fly. Keep in mind while reading further in the Bible, like the story of the flood, that God has seen all creatures and said the existence of all animals was good.

"And God said, Let the earth bring forth the living creature after his kind, cattle, and creeping thing, and beast of the earth after his kind: and it was so. And God made the beast of the earth after his kind, and cattle after their kind, and every thing that creepeth upon the earth after his kind: and God saw that it was good. And God said, Let us make man in our image, after our likeness: and let them have dominion over the fish of the sea, and over the fowl of the air, and over the cattle, and over all the earth, and over every creeping thing that creepeth upon the earth." Reading ahead a bit, the creation of man was on the sixth day so the animal creation must have been crammed into the fifth day as well as the fish and birds. Here the editor got a little out of synchronization with the pattern they established in listing the action and then giving it a date. In Genesis 1:21 Gods creates animals, in Genesis 1:22 He commands they be fruitful and multiply, and in Genesis 1:23 declares this was the fifth day. Then the story reiterates the creation of animals as prelude to the creation of man and woman. Following the flow of the story, God gives man dominion over the animals and this may be the need to retell the creation of the animals. The story flow then ends the sixth day on Genesis 1:31 leaving the impression the animals were created on the fifth day and retold on the sixth day when humans were added to the earth's living creatures. The writers were trying to say that God created the animals they way that they existed at the time of writing. In these lines the writers stated that God created the beast (and lets use a bear as an example) after his kind. This clearly states that a bear was created like bear. It did not evolve from a deer, or even a fish or bird that was in existence before God created the bear. In an awkward way this states animals did not evolve from change, they were created in their own likeness. This became the main argument against evolution, it clearly states God created animals the way that they exist. The writers and editors of the scriptures did not have knowledge of species that were isolated on islands or the differences they developed from same species on continents with differing environments, but this study is left for other biological sciences. Again, God has overlooked an exceptional opportunity to create creditability for the writers and editors of the Tanakh (the first five books of the Bible). Even a brief description of the existence of the dinosaurs, woolly mammoths and saber-toothed tigers inserted here would deflate many arguments for evolution. The writers and editors of the Tanakh had no knowledge of these creatures. Not including a description of such animals supports the theory that the story of creation was limited to the knowledge of the writers with no divine guidance. This point in the story would have been a great time to include descriptions of island birds and insects that differ from their continental relatives. The fact that God created these differing island birds and insects that differ from the continental relatives could have been told as God's great prank to trick evolutionists. Revealing the prank before the evolutionists discovered it would add creditability to the Bible and creation story. Its omission tends to reverse the creditability to the evolutionists over the creationists by showing a lack of knowledge by the Canaan writers. The absence of anything is not proof that it did not exist. The absence of detailed explanations and listings of animals does not prove the Tarah (or Tanakh or Pentateuch, the first five books that became the Bible) were not the inerrant word of God, it just proves there was an opportunity missed to prove they were the inerrant word of God. Several arguments may be made to explain the missing details like the writers were trying to minimize the length or focus on the main story. These arguments to defend the absence of detail present a paradox. The arguments are based on the fact that the scriptures were written by people to be short, concise, and focused to enforce the idea that God generated them. This paradox does not prove the absence of God's involvement but it shows a high probability of being written by people for people of the era in which it was written.

There is low probability that God would have written these books with the knowledge limited to the human understanding at the time. It seems probable that God would have included expanded knowledge for the people of the time and for future generations. Absence of extended knowledge is not proof of non-involvement but it is certainly an opportunity missed to demonstrate God's involvement.

God said "Let us make man in our image, after our likeness." Note God said, "US" and "OUR". God using a plural case of pronouns is significant. If this was a translation error, why was it not corrected through many of the translations of Aramaic, Latin, Greek, German or English? The use of plural pronouns may be a translation error but it is more likely to be an attribute of the original mythologies of Mesopotamia and Egypt that were polytheistic influencing the creation story. There is evidence that the Jewish religion was not always monotheistic. There have been fertility Gods and Idols from around 1000 BCE found in excavation sights now in Israel. The findings of these Gods and Idols indicate that during the time of the initial writing of the Bible, the Jewish religion was polytheistic.

There is no scientific way to determine the Bible's origin or the creator or creators. Parts of the book of Genesis (the first of the five books attributed to Moses, Pentateuch or Torah) were found in five of the 11 caves in Qumran, caves 1, 2, 4, 6, and 8. A brief description of the Qumran caves and community is provided in the Glossary and it is suggested as a subject for investigation. Unfortunately, Genesis, the first book was not fully recovered and the direct translation from Aramaic or Greek or Hebrew has not been made from the Dead Sea Scrolls. It is very likely that the first book was the outside of each scroll that contained it and was the least protected, therefore least likely to be fully recovered. No evidence exists of the original writings other than translations of the Bibles. The translations are consistent in the use of plural pronouns referring to God. The original authors are thought to be of the Jewish religion and the Jewish religion is monotheistic. The use of plural pronouns is consistent and yet contradictory to the religion for which the Bible was written.

The highest probability is that these descriptions were copied from polytheistic religions.

"So God created man in his own image, in the image of God created he him, male and female created he them. And God blessed them, and God said unto them, Be fruitful, and multiply, and replenish the earth, and subdue it: and have dominion over the fish of the sea, and over the fowl of the air, and over every living thing that moveth upon the earth. And God said, Behold, I have given you every herb bearing seed, which is upon the face of all the earth, and every tree, in the which is the fruit of a tree yielding seed, to you it shall be for meat. And to every beast of the earth, and to every fowl of the air, and to every thing that creepeth upon the earth, wherein there is life, I have given every green herb for meat: and it was so. And God saw every thing that he had made, and, behold, it was very good. And the evening and the morning were the sixth day." God created male and female in his own image, which is the image of God. The method of creation is important here as well as the simultaneous creation of male and female in God's image. Keep in mind the style of writing to compare with following chapters in Genesis. Also, the number of people existing on earth after this creation is two. This completes the sixth day so all that is left in the creation is the seventh day for rest.

God saw the creation and judged it as very good. Man and woman were created in the image of God, that would imply they had conscious decision-making capability. We know there is later a flood to destroy this creation God judges as very good. If society turned bad from very good there had to be reasons. If it were because of lack of education, guidance, and leadership, where were people expected to gain this education, guidance, and leadership? If the causes were not from lack of education, guidance, and leadership, then they had to be internal traits and instincts that God could not perceive when judging His creation as good. The dilemma is created with this judgment of everything as good and later in the story the world must be destroyed including animals with humans.

The verse 1:28 God has commanded man to be fruitful and multiply and replenish the earth. Replenish seems to be a strange command for populating a new world. This again reduces the probability of the Bible being the inerrant word of God.

If we consider God as benevolent, then his command to subdue the earth and all animals and have dominion over them also commands people to be caretakers to the earth and animals. The present global conditions and rate of elimination of species indicates we are not following this commandment from God. Is God angry? Religious leaders decree God is annoyed by our life styles but they do not assert that God is aware of our disregard for the extinction of animals or destruction of the environment. Religious leaders are quick to refer to AIDS as the wrath of God for lifestyle choices but there is no actual evidence supporting their opinion. Religious leaders claim earth quakes, tsunamis, hurricanes, volcanoes and natural disasters are God's wrath for human's disobedience. AIDS or natural disasters rarely touch the people in leadership positions that control the destruction of earth and animals. Natural disasters inflict the damage on the poor and working class of people. AIDS is a decease spread through live-cell viral contact. Stating these are a punishment is stating that God punishes the innocent and rewards the culpable. There is no evidence of God's punishment for destruction of the environment.

The last few verses before concluding the sixth day gives the impression all is good and God has provided fruits, vegetables, grains and herbs for human consumption and states twice that these are for meat to humans. These verses is may be read as a commandment for vegetarian lifestyle. God's commandment to use plants as meat for our consumption has not been followed well.

Questions over Chapter 3:

1. God created birds, fish, and mammals after his own kind means
    a. they looked like God.
    b. they looked like what they are or the way that the writers saw them.
    c. God knew what they were supposed to look like.

d. none of the above.

2. God created every animal, bird and fish
    a. the way they are today.
    b. the way they were at the time of writing the Bible, but they may have evolved since then.
    c. the way they were perceived by the Bible's authors who were unaware of species in the rest of the world.
    d. and they are unchanged.
3. The Dead Sea Scrolls do not verify the first books of Genesis
    a. because they are different.
    b. because Genesis was not in the Dead Sea Scrolls.
    c. because the first book of Genesis has not been fully recovered and is not likely to ever be fully recovered.
    d. because the Bible was translated and the Dead Sea Scrolls are in original Aramaic and Hebrew.
4. God created man and woman in his own image
    a. by commanding it.
    b. by scooping up clay and forming them.
    c. by ordering other gods to create them.
    d. we are not told why, but it appears probable that the writers had to designate humans as god-like to distinguish humans as above other animals.
    e. is likely an error in translation, and the actual case was that man created God is his own image.
5. The plural pronouns of 'us' and 'our' were used
    a. to indicate man was to look like male gods and woman to look like female gods.
    b. to get other gods to help Him.
    c. due to a translation error.
    d. because it was probably a from polytheistic religions.

6. God judged the creation on the 6th day as very good because
    a. no sin had been committed.
    b. everything was in harmony and there were no problems.
    c. the world was set as a paradise.
    d. of any or all the above because we are not given the reason.

Probability of statements in Chapter 3:

Instructions: Assign a number to the plausibility of each statement.
(1) For a total contradiction to a statement.
(2) For an implausible statement.
(3) For a possible and maybe even practical statement.
(4) For a plausible statement.
(5) For a confirmed or verifiable statement.

1. God saw everything He had made and said it was very good. What is the probability that nothing has ever changed from original creation? (1 – 5)?
2. What is the probability that the plural pronouns are due to a translation error and have nothing to with polytheistic religions? (1 – 5)?
3. What is the probability that God created all animals and made small variations between those that are on islands and those on the main lands? (1 – 5)?
4. The flow of the creation of animals and man is ambiguous at best. It may be read that animals were created on the fifth day and man on the sixth day or that man and animals were created on the same day and not clear if that was the fifth, sixth or spread over both. This appears to be ancient legends over the thoughts of a super powerful being writing. What is the probability that this supports the idea that the Bible is the inerrant word of God? (1 – 5)?
5. Selective breeding by humans has generated present day dogs' functions--companionship, loyalty, and size, and many other characteristics. What is the probability that God did not create these traits in dogs because He

did not want humans to have dogs as they exist today? (1 – 5)?

6. God commanded man to have dominion over fish, animals and every living thing, giving humans authority as well as responsibility. Regardless of religious believes and interpretation of this command, man became the ultimate predator. Writers of the Bible did not foresee the potential for destruction. What is the probability that this commandment was from super natural power? (1 – 5)?

Chapter 4: The Second Creation of Heaven and Earth

Book of Genesis Chapter 2

2:1 Thus the heavens and the earth were finished, and all the host of them.

2:2 And on the seventh day God ended his work which he had made, and he rested on the seventh day from all his work which he had made.

2:3 And God blessed the seventh day, and sanctified it: because that in it he had rested from all his work which God created and made.

2:4 These are the generations of the heavens and of the earth when they were created, in the day that the LORD God made the earth and the heavens,

2:5 And every plant of the field before it was in the earth, and every herb of the field before it grew: for the LORD God had not caused it to rain upon the earth, and there was not a man to till the ground.

2:6 But there went up a mist from the earth, and watered the whole face of the ground.

2:7 And the LORD God formed man of the dust of the ground, and breathed into his nostrils the breath of life, and man became a living soul.

2:8 And the LORD God planted a garden eastward in Eden, and there he put the man whom he had formed.

2:9 And out of the ground made the LORD God to grow every tree that is pleasant to the sight, and good for food, the tree of life also in the midst of the garden, and the tree of knowledge of good and evil.

Chapter 2 changes the tone of the story and also contradicts the first chapter. Chapter 1 is a direct narration of the actions and results of the creation. Chapter 2 is a more priestly account. The writers summarized the work and gave implied direction from God's actions. God 'sanctifies' the seventh day as a day of rest implying this should be followed by the human race. At this time there is no work to be done because man and woman as they are created will live in paradise. Why do they require a day of rest? They rest seven days a week because they live in paradise or else it is not paradise. The other factor showing the changing of authors from the first chapter is the contradictions. Heaven and earth were apparently finished on the sixth day, which is in agreement. Chapter 2 has God completing his task of creating the world and rested on the seventh day and there is no man. Chapter 1 states God created man (i.e. human beings) both male and female in his own image on the sixth day. Chapter 2 does not have humans existing until after the seventh day.

We have to assume the story is in the eighth day and nothing has started to grow since it has not rained. The plants and herbs are in the earth in chapter 2 but chapter 1 implies the plants and herbs were growing and yielding seeds. Since they have existed for one day the plants and herbs were likely to survive a few days without rain, so not raining yet was not a catastrophe. To ensure all vegetation continued a mist went up from the earth and watered the whole face of the earth. So this is a very minor difference if the Bible is written by ancient people but is it minor if it is the inerrant word of God?

Now God forms man out of dust of the ground. This is more descriptive than the creation of humans in the first chapter and certainly different. After forming man as a statue, God has to give him life by breathing into his nostrils. Many civilizations before the writing of the second chapter of Genesis had created images or statues of men and women. Having God form a statue and then breathing divine breath into the nostrils appears to be a convenient visualization for authors and scribes. The ancient mythologies brought life to new creations from parts of existing life, such as the blood of a deceased God causing life in other religions. Ancient mythologies limited the creations of life and instead passed it on from remains of deceased Gods. God breathing divine breath into the new man of dust follows this tradition that life has to begin from life. This scenario gives the limitations of the biological understanding of the writers of this chapter of Genesis. The writers equated life with breathing, and ignored blood flow, nerve excitation and the other biological functions required for life. The extreme over simplification of life maybe ignored but the blatant differences are much more difficult to ignore if the Bible is the inerrant word of God. The differences in the second chapter are: first there is only one human, a male, the creation of man is after the seventh day, not on the sixth, the second creation of man is not necessarily in God's image as was the first, and in the second creation man becomes a living soul which is not stated in the first creation.

For a special place God created a garden in eastward Eden. The "Garden of Eden" has never been found. Many centuries of searching have never produced any actual evidence that the Garden of Eden ever existed. The Bible does not give us the location of the Garden of Eden. But the Bible says it was there and God placed man in it. There is however some description of the Garden of Eden as it grew every tree that is pleasant to the sight. There was no mention of the Garden of Eden in the first chapter of Genesis, only in the second chapter. If the Garden of Eden has every tree that is pleasant to the sight, what a plethora of trees that would include: palms, fir, mesquite, giant redwoods, and thousands more. This does not say if the trees were only living or if the Garden of Eden included petrified logs from the Jurassic era as in the Arizona desert since they are pleasant to look at. The Garden of Eden may no longer exist since every tree that is pleasant to the sight would require different climates to live any length of time, excluding the petrified logs from the Jurassic era. The "and" in the description of the trees may be used in the sense of a logical 'and'. A logical 'and' requires that both conditions of a statement be true for the statement to be true. So, this could imply that trees in the garden were both pleasant to the sight AND good for food. The statement requiring both conditions be met would certainly reduce the number of different types of trees. Yielding good food to eat certainly eliminates the petrified logs. Still Euell Gibbons, the late authority on edible wild plants, found nutrition in pine nuts and seeds from northern trees as well as coconuts and tropical fruits from aesthetically pleasing trees. Even with the constraints of multiple restrictions imposed by the use of the logical 'and', the number of trees would be enormous and require climates not compatible with all species.

Turn now to the two specific trees mentioned. One is the tree of life and the other is the tree of knowledge of good and evil. These are not average mango and papaya trees. The verse of the third day in Chapter 1 tells us God brought forth all vegetation and trees yielding fruit. It seems a bit flippant for the inerrant word of God to not mention these trees in chapter 1 as part of the gargantuan selection of vegetation. Here the authors of the Torah scramble the order of information to stifle contiguous thought flow. Let's jump ahead a bit to continue the thoughts on these two trees. God tells the man not to eat of the tree of knowledge or he will die. Why was the tree of life not included in this death threat? It appears the authors drew a parody of the Egyptian mythology of Shu (air), Tefnut (moisture), and their offspring Nut (sky) and Jeb (earth). Shu learns of his children's unnaturally close relationship for siblings and separates them to create earth and sky. In Egyptian religion these four are gods and having four gods is not acceptable for a monotheistic religion. The insertion of the knowledge of sexual intercourse was too important of a message for the authors to omit, so gods were converted to a tree of knowledge to portray the message.

The tree of life has several connotations connected with it. The reference is vague, therefore, inferences the authors were trying to make are not clear. Beliefs existed in several mythologies predating the Torah that may have guided development of the Torah that a key to immortality existed but was obscure. It is possible that the belief in immortality was influenced from the Egyptian belief that the 'soul' or 'ba' was reunited with the body in the afterlife. This was the reason Egyptians put so much effort in preserving the body with several methods of mummification.

It is probable that to keep a source of immortal life for the afterlife and to ensure the religion to remain monotheistic, immortality was converted to the tree of life and added to the garden.

Questions over Chapter 4

1. God sanctified the seventh day
    a. so humans could rest.
    b. because He rested after his work.
    c. to make it a day for sports.
    d. to make it a day for worship.
2. God completed heaven and earth and rested on the seventh day
    a. but man and woman were not created yet so this differs from chapter1.

   b. man and woman rested also to please God.
   c. and God saw it was very good.
   d. God still had to create the stars.
3. God created all plants before resting on the seventh day
   a. and the plants started to bare fruit and multiply.
   b. and the earth was lush and green.
   c. and God made a mist rise to water the plants, apparently after the seventh day.
   d. and plants do not wilt after a few days so every thing was good.
4. The differences in the first and second chapters are
   a. in the second man is formed of dust and not necessarily in God's image, in the first man was created in God's image simultaneously with woman.
   b. in the second God breathed into man's nostrils to give man life and a soul, which was not mentioned in the first chapter.
   c. in the first God created man and woman on the 6th day, not the 8th day.
   d. all the above and several more.
5. God created man from dust
   a. because we return to dust.
   b. to mold man in his own image.
   c. to show his power.
   d. as a form to breathe life into his nostrils.
6. God planted the garden east of Eden
   a. which was lush with beautiful flowers.
   b. for Adam and Eve.
   c. and added every tree that was pleasant to look at and every tree that provided food and the tree of knowledge and tree of life.
   d. all the above.
7. God placed man in the Garden of Eden
   a. to be with the woman.
   b. but we arc not told why.
   c. to have access to the trees of knowledge and life.
   d. to share with God.

8. The tree of life and tree of Knowledge were placed in the Garden of Eden
    a. but we are not told why, but it seems to be a carryover from other religions.
    b. to tempt man.
    c. as a home for snakes.
    d. to complete the garden because there were not enough trees.

Probability of statements in Chapter 4

Instructions: Assign a number to the plausibility of each statement.
(1) For a total contradiction to a statement.
(2) For an implausible statement.
(3) For a possible and maybe even practical statement.
(4) For a plausible statement.
(5) For a confirmed or verifiable statement.

1. Statues being brought to life existed in other mythologies before the Bible was written. Such myths may have influenced the writing of the second creation. What is the probability that both creations came from one God? (1 – 5)?
2. Two different creations with no explanation for the differences make them contradictions. What is the probability that an inerrant document may have contradictions? (1 – 5)?
3. God rests on the seventh day, which becomes a law for the Jewish religion. The law for no work on the Sabbath has continually diminished in importance with increasing technology that cannot shut down for a day. What is the probability that God's law of resting on the Sabbath and doing NO work changed with technology? (1 – 5)?
4. What is the probability that the Garden of Eden could sustain every tree that was pleasant to look at and every tree that provided food in one climate? (1 – 5)?
5. The tree that provided knowledge of good and evil and the tree that provided immortality are very similar to Egyptian mythology's Nut (sky) and Jeb (earth) who

instruct morality and provide immortal life. What is the probability that the Bible was written with no knowledge of the Egyptian religion and results in a similar principle? (1 – 5)?

6. Many factors tend to suggest the Bible stories were not the inerrant word of God, the overall simplification of the creation, lack of technical understanding, the similarities to other religions and mythologies, self-contradictions, and contradictions with reality. What is the probability of divine guidance in the writing of these stories in the Bible? (1 – 5)?

Chapter 5: Locating the Garden of Eden

Book of Genesis Chapter 2

2:10 And a river went out of Eden to water the garden, and from thence it was parted, and became into four heads.
2:11 The name of the first is Pison: that is it which compasseth the whole land of Havilah, where there is gold,
2:12 And the gold of that land is good: there is bdellium and the onyx stone.
2:13 And the name of the second river is Gihon: the same is it that compasseth the whole land of Ethiopia.
2:14 And the name of the third river is Hiddekel: that is it which goeth toward the east of Assyria. And the fourth river is Euphrates.
2:15 And the LORD God took the man, and put him into the garden of Eden to dress it and to keep it.
2:16 And the LORD God commanded the man, saying, Of every tree of the garden thou mayest freely eat:
2:17 But of the tree of the knowledge of good and evil, thou shalt not eat of it: for in the day that thou eatest thereof thou shalt surely die.

Picking up again with our creation story we find the disorganizing of data to dazzle our logical thought process has continued. The authors wish to give us the location of the Garden of Eden. "And a river went out of Eden to water the garden, and from thence it was parted, and became into four heads." Hundreds of years of searches by a multitude of scholars have not discovered the location of the Garden of Eden. The impression received from reading this chapter of Genesis is that the authors were disclosing the location. Eden's location has been narrowed down to the triangle from North Central Africa to India to southeastern Europe. While this eliminates ninety percent of the earth, it was the entire known world to the authors of the Torah. The absence of location of Eden or the Garden of Eden creates the question was Eden an actual place or the imaginary ideal paradise to enhance the creation story. The desire to add creditability to the Bible has driven the search for Eden. Biblical scholars failing in this search for Eden, claim the description provided was too elaborate to be added for robustness to an imaginary fable, therefore The Garden of Eden must have existed. This is a science textbook and the conflict of the existence of Eden is resolved through factual evidence and removing rhetoric.

First is the description to intricate for a fable? No. This is not evidence of existence of the Garden of Eden. Many detailed and intricate descriptions are found in fictional stories, such as the Harry Potter stories or Stephen King novels. The burden of writing is greatly reduced for modern authors over scribes from 2000 to 3000 years ago. However, this reduction in difficulty in writing is in the physical transfer of thoughts to media such as papyrus, parchment, leather or paper. The act of creating thoughts may also be somewhat easier in that today authors have much greater exposure to broader horizons of knowledge than several millenniums past. This does not ensure a restraint that writing would have been such a burdensome task as to confine writing to honest disclosure of divine communiqué and eliminate the realm of creative writing. Therefore, the argument that Genesis 2:10 through 2:14 is too complex to be imaginary is not true.

Second where is Eden? Science cannot prove that something never existed. With no evidence found after millenniums of searching by multitudes of scholars, it seems most probable that it did not exist, but gives the story the benefit of inconclusiveness[3]. In the time of writing the Torah, lush areas would have existed in Arabia, Egypt, Ethiopia, Mesopotamia, India, Babylonia, Assyria[4], etc. that could have inspired the writers to image a paradise for the birth of the human race. Locating Eden then has to be estimated by highest probability and starting with deciphering the names of the rivers and lands. Describing the location of Eden by the rivers shows the allegoric nature written in the creation story. At the time of writing the Torah, part of the Jewish population had returned from Egypt and Assyria or Babylon. Social and economic life was dependent on the rivers. The Nile, Euphrates, and Tigris rivers were essential to development of the areas and civilization to this point in history. The rivers' importance to social life at this time then guides the authors to using rivers as focus for Eden and the garden setting for paradise.

The river Pison is unknown. Speculation defines it as a tributary to either the Nile or the Euphrates. (Note that the

Nile is in Egypt and the Euphrates is in Iraq.) There exist dry riverbeds that were tributaries to the Tigris and Euphrates at one time or which flowed only during heavy rains. The heavy rains creating these now dry riverbeds are now rare but changes in landscape indicate they were more common several millenniums past. It is interesting to note that the authors used the present tense of verbs in writing this story of creation. This would indicate the river was not a dry bed but a water flowing river. This complicates the search for Eden even more as the Sahara and Arabian Deserts existed during the writing of this story and rivers have changed little since then. It could be that this is part of the allegory that the writers were writing in present tense of what they believed the landscape was like during the time frame they set for the creation and imply this landscape was lost with the Garden of Eden.

The verse states the river Pison is in Havilah. Havilah is not known. Guesses are that it is part of northern Arabia. This would be consistent with Pison being a dried tributary to the Euphrates. Some scholars have translated this to Hiyal or Ha'ill to place it in central northern Arabia. A problem arises in defining the land as having good gold, bdellium, and onyx stone. For readers unfamiliar with bdellium, it is an aromatic gum resin like myrrh that was exuded from trees in Havilah. Bdellium may have been popular at the time of writing the Torah but not a big seller today. Onyx stone also may have been in demand a few thousand years ago but not the stuff to put your city on the map. But gold has seemed to hold its luster through generations. If there were gold in Havilah it seems likely it would have been a very popular location and attracted a large population. The lack of a reputation does not prove there was no gold and that Havilah did not exist. However, the lack of any knowledge of the location of Havilah and the statement that it did have gold "and the gold of that land is good" does reduce the probability of this being a factual account over an allegory.

The second river Gihon is again unknown. Locating this river is even more difficult because it is said to encompass Ethiopia. The river Gihon has led people to speculate Eden could be in North Africa. Ethiopia today is the source for the Nile. Ethiopia is on the eastside entrance the Red Sea and over

1500 miles from Egypt. For Gihon to intersect the other rivers that are in the story, it has to get north of Babylon or through Assyria and cross the Red Sea and the Persian Gulf and exceed the length of the Nile. The Gihon did not do this. The commonly accepted explanation with no verification is the name 'Ethiopia' was confused due to interchanging the names of a nationality and the name of a country. Ethiopia translated from Hebrew is 'Cush'. During the time of writing the Torah, people were named after a ruler as well as an area.

Book of Genesis Chapter 10 (excerpt for information)

10:8 And Cush begat Nimrod: he began to be a mighty one in the earth.
10:9 He was a mighty hunter before the LORD: wherefore it is said, Even as Nimrod the mighty hunter before the LORD.
10:10 And the beginning of his kingdom was Babel, and Erech, and Accad, and Calneh, in the land of Shinar.

The Bible says Cush was a King or ruler that fathered Nimrod, that eventually ruled areas near Babylon and in Mesopotamia where the river Gihon could have been located. This area or land ruled by Cush would have Cush (translated to Ethiopia) for the name of the land. It is possible that the people of this area may have been referred to as Cushites or translated to Ethiopians. Thus, Cush or Ethiopia was located in the upper Mesopotamian Plain where the river may connect with the Euphrates or Tigris to be one of the four rivers. This scenario of adapting the local area name and translating it to Ethiopia is much more plausible to be copied into the story of Eden than the actual location of Ethiopia.

The third river is Hiddekel, which is translated to Tigris, and the fourth river is Euphrates. The Tigris and Euphrates combine just before dumping into the Persian Gulf in Mesopotamia. This land is now called Iraq. The Euphrates River runs diagonally from the northwest to the southeast near the middle of Iraq, and the Tigris is east of the Euphrates. These two well-known rivers suggest the location of Eden and the two obscure rivers were in Mesopotamia and the description of Eden was brought back from Babylon with the Jews returning from exile. The most probable location the writers were using would be in what is now Iraq. No evidence of the "Garden of Eden" suggests this is a parable or allegory.

God placed "man" in the Garden of Eden. Until this point in the Bible, the Bible only mentions God as creating one man, we must assume this reference is to one man and not man as plural, or referring to mankind. This man is placed there to be the gardener to keep the Garden. God gives man one order, he must not eat of the tree of knowledge or he will die the day he eats of the tree of knowledge. God did not forbid the man from eating of the tree of life. What if the man had eaten from the tree of life and then from the tree of knowledge? Which tree had more power, the tree to sustain life or the tree of knowledge, which was told to be deadly?

The Egyptian Heliopolitan Creation from Egyptian Coffin text 80 contains an extensive philosophical presentation of immortality and moral behavior[5]. The Egyptian mythology gives Shu (air) and Tefnut (moisture) as the offspring of or Re-Atum or Atum. In the Egyptian mythology Shu also represents the deity for life or immortality and Tefnut represents the deity for moral behavior or the knowledge of good and evil. The Jewish religion was monotheistic and could not include Shu and Tefnut as deities and substituted trees baring fruit of the same purpose. Including Egyptian deities in altered forms to eliminate the Godly status is a consistent practice in the Biblical creation. Later God calls the man 'Adam'[6]. Adam seems to be orally close to Atum the first God in the Egyptian Religion. Again, calling man Adam is not proof of plagiarism, but it increases the probability of being copied from Egyptian mythology.

Questions over Chapter 5

1. Where was the Garden of Eden?
    a. Eastward of Eden.
    b. In any of the known world to the writers.
    c. It is not known.
    d. All the above.
2. Did the Garden of Eden actually exist?
    a. This cannot be proven.

   b. Yes, because the description is too complicated to be fabricated.
   c. No, because it cannot be found.
   d. Yes, because it is the beginning of humans.
3. Are the four rivers real?
   a. Yes, but translations loss their names.
   b. This cannot be determined, but it seems likely to be a fantasy of the writers.
   c. No, because they cannot be located.
   d. Not today but they were rivers at one time.

4. Could it be that Eden exists and the translations obscure it?
   a. Yes, because we know Cush could mean two different locations.
   b. Yes, because the rivers make sense.
   c. It is highly improbable because of hundreds of years of searching have not located Eden.
   d. No, because we know how to translate now.
5. God put the man in the Garden of Eden
   a. to be with Eve.
   b. to start the human race.
   c. as a care taker.
   d. to protect him.
6. God told man not to eat of the tree of knowledge of good and evil
   a. but did not mention the tree of life.
   b. or he would die that day.
   c. because this tree and the tree of life appear to be fragments of polytheistic religions brought back from Egypt and Mesopotamia.
   d. all the above.
7. The trees of life and knowledge of good and evil
   a. were unique to the Garden of Eden.
   b. are similar to Egyptian Heliopolitan Creation from Egyptian Coffin text 80.
   c. were forbidden.
   d. all the above.

Probability of statements in Chapter 5

Instructions: Assign a number to the plausibility of each statement.
(1) For a total contradiction to a statement.
(2) For an implausible statement.
(3) For a possible and maybe even practical statement.
(4) For a plausible statement.
(5) For a confirmed or verifiable statement.

1. What is the probability that the four rivers that locate the Garden of Eden can just be language translation errors? (1 – 5)?
2. What is the probability that the descriptions of the location of Eden are too complex to be fabricated and this assures the story to be true? (1 – 5)?
3. Naming kingdoms and areas after rulers may complicate locating the rivers that define the location of Eden. What is the probability that this proves it had to exist but cannot be found because of translations. (1 – 5)?
4. God added the tree of knowledge to temp man because he would die with in the day of eating from it. What is the probability that this is different from putting out rodent bait? (1 – 5)?
5. God placed man in the Garden of Eden as a caretaker. The story does not mention any tools like axes, shovels, mowers, trimmers, etc. The job of caretaker must be been pretty tough. What is the probability that an unequipped man can care for the Garden of Eden? (1 – 5)?
6. Havilah is said to have gold "and the gold of that land is good". Is it likely that a place having gold would have no evidence of ever existing? (1 – 5)?

Chapter 6: The Second Story of Populating the Earth

Book of Genesis Chapter 2

2:18 And the LORD God said, It is not good that the man should be alone, I will make him an help meet for him.
2:19 And out of the ground the LORD God formed every beast of the field, and every fowl of the air, and brought them unto Adam to see what he would call them: and whatsoever Adam called every living creature, that was the name thereof.
2:20 And Adam gave names to all cattle, and to the fowl of the air, and to every beast of the field, but for Adam there was not found an help meet for him.
2:21 And the LORD God caused a deep sleep to fall upon Adam, and he slept: and he took one of his ribs, and closed up the flesh instead thereof,
2:22 And the rib, which the LORD God had taken from man, made he a woman, and brought her unto the man.
2:23 And Adam said, This is now bone of my bones, and flesh of my flesh: she shall be called Woman, because she was taken out of Man.
2:24 Therefore shall a man leave his father and his mother, and shall cleave unto his wife: and they shall be one flesh.
2:25 And they were both naked, the man and his wife, and were not ashamed.

The first story of the creation in the Book of Genesis 1, God created all animals and sea creatures on the sixth day and then created a man and woman. The second story God creates a man, let's call him Adam, then the animals. God lets the man Adam name the animals. Lucky for us, Adam knew what electricity was to name the electric eel and what neon light color was to name the neon fish. This cannot be. Adam did not know of the existence of all living creatures and thus could not have named them. Galapagos Islands, Madagascar, New Zealand and the Americas had living creatures of which Adam had no name or even knowledge of their existence.

Adam found no help in naming the animals. Maybe this statement indicates that Adam required assistance in surviving or living not in actually naming the animals. From the next line it appears the LORD God understood this and decided to create a spouse for Adam. This line Genesis 2:21

shows the writer of this creation story wants to separate humans from animals. Animals were created out of ground, i.e. earth or dust. Here the woman is created from one of Adam's ribs. This story of creating man and woman are different from the first story where God created both from dust in his own image and likeness. The next line 2:23 sets the precedence that woman is a subset of man. Is this to initiate the argument that woman is derived from man and thus inferior to man. This has been used as arguments for women to be subservient to men. Genesis 2:23 is Adam speaking and there is no change in verse 24 to change our belief that Adam is continuing to speak to the reader. Now the man, Adam was just created or born yesterday and the woman was just created, and Adam explains the structure of marriage. Seems he may have already eaten from the tree of knowledge of good and evil to gain such in depth knowledge of life after just being born yesterday.

Adam could not have known what was to be expected of life. Adam had no father and mother. So the possibilities are that this is God explaining to Adam the structure of life and marriage or this is the redactors (editors) inserting their opinion of life. To this point the scribes and redactors have been vigilant to denote quotes from God and specify that God said 'etc'. Genesis 2:24 is not given that God said, "Therefore shall a man leave his father and his mother, and shall cleave unto his wife: and they shall be one flesh" therefore it has be the scribes and redactors inserting their opinion. If the scribes and redactors are inserting their opinion, then is the Bible the inerrant word of God?

The man and woman were naked and felt no shame. Hey, come on they were born yesterday, they were the only two people on earth, how could they develop inhibitions? The point is being made that the man and woman, Adam and Eve had no inhibitions or were naïve.

Questions over Chapter 6

1. God created all animals to
    a. help man.
    b. because man should not be alone.
    c. to give man something to name.
    d. all the above.
2. The first man could not have known all animals
    a. so 'every' is an exaggeration.
    b. this only applies to animals that were known.
    c. to show the first man even though created for only one day had full intelligence.
    d. was stated this way to make the story richer.
3. God put the man to sleep and took a rib to create a woman
    a. which is why men have one less rib than women.
    b. because it would have hurt if man was awake.
    c. which is why in general women are smaller than men.
    d. which is different from the first creation where God creates both in his image.
4. God created the first woman from the man's rib
    a. to show women are a subset taken from men but still equal.
    b. to show women to be equal to men.
    c. because taking life from life was a concept similar to other religions.
    d. and therefore is called woman to mean man's rib.
5. Adam names the woman and goes on to preach the virtues of a monogamous relationship
    a. because Adam had to leave his parents to take a wife.
    b. because Adam knew the woman was part of him.
    c. which shows a deep understanding of relationships like leaving his parents and creating a stable marriage with NO experience.
    d. because Adam wants to teach us about life.

6. The lack of inhibitions about nudity
    a. was because they were the only people on earth.
    b. because they had no reason to be ashamed.
    c. which seems contradictory to the deep knowledge of marriage and life.
    d. all the above.

Probability of statements in Chapter 6

Instructions: Assign a number to the plausibility of each statement.
(1) For a total contradiction to a statement.
(2) For an implausible statement.
(3) For a possible and maybe even practical statement.
(4) For a plausible statement.
(5) For a confirmed or verifiable statement.

1. What is the probability that Adam and Eve neither had parents but knew that they were to leave their parents for marriage? (1 – 5)?
2. What is the probability that God removed a rib from Adam and the probability that this explains why men have one less rib than women? (Men and women both have 24 ribs, 12 on each side). (1 – 5)?
3. What is the probability that since women are generally smaller in size than men, that the Bible is correct in suggesting that they are a subset and implies a lower level or status? (1 – 5)?
4. What is the probability that Adam named all animals? (1 – 5)?
5. What is the probability that the writers created this story independent of other mythologies that were based on life being only drawn from life? (1 – 5)?
6. What is the probability that Adam and Eve knew enough about marriage to direct monogamy when there were only TWO humans. (1 – 5)?

Chapter 7: The original Sin

Book of Genesis Chapter 3

3:1 Now the serpent was more subtil than any beast of the field which the LORD God had made. And he said unto the woman, Yea, hath God said, Ye shall not eat of every tree of the garden?
3:2 And the woman said unto the serpent, We may eat of the fruit of the trees of the garden:
3:3 But of the fruit of the tree which is in the midst of the garden, God hath said, Ye shall not eat of it, neither shall ye touch it, lest ye die.
3:4 And the serpent said unto the woman, Ye shall not surely die:
3:5 For God doth know that in the day ye eat thereof, then your eyes shall be opened, and ye shall be as Gods, knowing good and evil.
3:6 And when the woman saw that the tree was good for food, and that it was pleasant to the eyes, and a tree to be desired to make one wise, she took of the fruit thereof, and did eat, and gave also unto her husband with her, and he did eat.
3:7 And the eyes of them both were opened, and they knew that they were naked, and they sewed fig leaves together, and made themselves aprons.

Adam and Eve are in the Garden of Eden. The general interpretation of the serpent is that it is a large snake. The serpent speaks to Eve. Eve is not alarmed at a talking snake since she has no knowledge that a snake does not have voice capability. Animals do communicate without a larynx to give warnings and directions of food, water, shelter, or safety. Some birds may actually imitate the human voice. Animals do not communicate decently with human type communication. Chimpanzees have been taught sign language to substitute for absence of voice to enable communication but this is after years of training and with less sophistication than required for the communication the serpent is reported to have given to Eve.

Adam and Eve have developed a good communication. God told Adam not to eat of the tree of knowledge and God has not told Eve, yet Eve knows God has commanded this. Actually Eve has information beyond what God has given to Adam in that she says she may not even touch the tree. The serpent tells Eve she will not die and the serpent gives Eve even more information. The serpent says that the day she eats of the tree of knowledge, her eyes will open and she will be as Gods, knowing good and evil. Note that the serpent uses the plural of Gods, indicating in excess of one.

We are not told the time lapse between being created and Eve conversing with the serpent but the lack of it being expressed would indicate a relatively short period of time. After careful examination, as much as may be expected after the experience gained from a short time of living, Eve decides to eat the fruit of the tree of knowledge. Then the woman, Eve gives it to her husband, Adam. Adam does not question eating this fruit that he has been told by God will certainly result in death the same day that he eats. We have to assume like all good husbands if his wife has eaten then he should eat and they go together or live together with increased knowledge.

Now their eyes were open indicating the man and woman, Adam and Eve had gain knowledge of good and evil. They did not die, so God lied to Adam. Adam and Eve became aware of their nakedness. Were they alone in the Garden of Eden except God? Why would sudden cognizance of their lack of dress code compliance embarrass them? How does having knowledge of good and evil suddenly make being naked wrong or evil? It appears the writers are adding their opinion of good and evil and considering nakedness as evil.

They sewed fig leaves together to make aprons for minimum cover. Apparently, the tree of knowledge provided more than just knowledge of good and evil, as now they were as resourceful as the professor on Gilligan's Island. Enough for flippancy, this is serious. This is considered the original sin by Jewish and Christian religions. The sin here is a carryover from Egyptian and Mesopotamian (Assyrian and Babylonian) religions in the belief that knowledge was reserved for the Gods. The point stressed here is that knowledge made Adam and Eve cognizant of themselves and their world. The authors of the Bible are developing a case for the non-existence of the prefect utopia or the Garden of Eden. If man had no understanding of good and evil or right and wrong, then he could not sin. If we speculate on the writers' implication, there is no need for any sin because everything is provided for human needs and there is no need for steeling, killing, or the struggle for life that existed during the writing of Genesis. The story has to lead to the present life, as the writers knew during creating Genesis. Also the story has to explain the absence of utopia and the existence of sin, so partaking of the tree of knowledge completes the task. But if we take the story of creation as true, then we have to ask why would God place the tree of knowledge and the tree of life in the garden for temptation? Was God vindictive and wanted to provide utopia to Adam and Eve and take it away. Why would God remove the tree of knowledge of good and evil after He found that Adam and Eve did not become Gods after eating? If God is benevolent then it is contradictory for God not allow the tree of knowledge of good and evil to exist out side the Garden of Eden. The knowledge of good and evil would be an asset to society possessing so many political leaders, evangelist and religious leaders and to have them partake of its fruit.

Book of Genesis Chapter 3

3:8 And they heard the voice of the LORD God walking in the garden in the cool of the day: and Adam and his wife hid themselves from the presence of the LORD God amongst the trees of the garden.
3:9 And the LORD God called unto Adam, and said unto him, Where art thou?
3:10 And he said, I heard thy voice in the garden, and I was afraid, because I was naked, and I hid myself.
3:11 And he said, Who told thee that thou wast naked? Hast thou eaten of the tree, whereof I commanded thee that thou shouldest not eat?
3:12 And the man said, The woman whom thou gavest to be with me, she gave me of the tree, and I did eat.
3:13 And the LORD God said unto the woman, What is this that thou hast done? And the woman said, The serpent beguiled me, and I did eat.

The LORD God is walking through the Garden talking or singing and Adam and Eve hide from Him. The deity that created the earth, the solar system and the universe and then created all life on earth, God, and He is walking through the garden. To avoid contact with God, Adam and Eve hide behind trees. God calls Adam out and Adam shows his fear and reveals his awareness of his situation. The only two people on earth are hiding like children playing hide and seek, from God. From Adam's statement that he was naked, we have to assume that an apron of fig leaves is socially unacceptable to appear before God. God appears near human level, although fairly perceptive. God ask Adam, "Who told thee that thou wast naked?" Has God lost track of the number of people he has created, or was this just an intimidating lead into questioning if they had eaten of the tree of knowledge? God may have needed to ask since there were talking animals in the garden and perhaps one of the animals was smart enough to have revealed to Adam and Eve their condition, which was being naked. Animals have never worn clothing except pets that people have dressed, but it is safe to assume there are no dressed pets in the Garden. So, the talking animals would have been completely undressed and not been likely to have told Adam and Eve that they were undressed. God would have known this so his second question was rhetorical for intimidation of Adam. God asked Adam if he had eaten of the tree. Adam not wanting to press his luck on this certain death consequence of eating from the tree God had forbidden, decides to share the blame with Eve. So, we know Adam has learned a little about being a husband even though he gave the speech on its virtues. God directs his inquisition to Eve. The scripture has not told us that God warned Eve of eating from the tree. She does know of the warning, because she protested when the serpent offered the fruit and she quotes God as the source of the information. Eve blames the serpent for convincing her the penalty was bogus and the rewards would be knowledge making them equal to the Gods for eating of the forbidden fruit. Neither Adam nor Eve used the excuse they were just born yesterday and were still naïve. Neither Adam nor Eve question God as to why He lied to Adam about the certain death on the day of eating the fruit. If they had eaten of the tree of life prior to the tree of knowledge, would they have achieved immortality? The story does not tell us that they died the very day that they ate of the forbidden fruit. So, God has lied to Adam and assumed Eve, and yet God is chastising them for eating of the fruit about which He lied. This does not appear to be the word of God when the story discredits God as dishonest and tormenting.

Questions over Chapter 7

1. God is strolling through the Garden of Eden in the cool of the day and
    a. calls out to Adam.
    b. cannot see Adam and Eve because they hide behind trees.
    c. is angry because Adam and Eve have eaten of the tree of knowledge.
    d. enjoying Adam and Eve's company because the cool of the day is morning.
2. Eve trusted a serpent with her life
    a. because it could talk so it must have been a rather intelligent serpent.
    b. because it tempted her and it shows humans give into temptation.
    c. because she was naïve and had no experience.
    d. any of the above could fit, as this was just the introduction to establish the sin that religious leaders now call the original sin.
3. Eve actually ate because
    a. the serpent told her she would not die and she would gain knowledge as the Gods.
    b. because there was a desire for knowledge even though the fruit looked undesirable but the tree was pleasant to look at.
    c. Adam agreed with the serpent.
    d. we are not told.
4. Eve ate of the tree of knowledge and then gave fruit to Adam
    a. so they would die together.
    b. because sharing was natural for only two people.
    c. and they both became knowledgeable as we are told by the use of their eyes being opened.
    d. because the serpent said to give it to Adam also.
5. Adam and Eve now having knowledge
    a. were ashamed of their nakedness.

b. put on fig aprons God provided.
c. still had very little taste in dress.
d. changed nothing.

6. Adam and Eve now having knowledge
    a. could talk intelligently with God.
    b. hid from God.
    c. were proud of their new gained knowledge.
    d. all the above.
7. Adam and Eve ate only from the tree of knowledge
    a. because that is all the serpent suggested.
    b. because the tree of Life was forbidden.
    c. because eating from the Tree of Life would make them immortal and not followed the story line the writers wanted.
    d. we are not told but is seems likely that it would not align with the other religions.
8. After eating the fruit Adam and Eve
    a. did not die.
    b. became mortal and would die.
    c. knew enough to confront God.
    d. all the above.

Probability of statements in Chapter 7

Instructions: Assign a number to the plausibility of each statement.
(1) For a total contradiction to a statement.
(2) For an implausible statement.
(3) For a possible and maybe even practical statement.
(4) For a plausible statement.
(5) For a confirmed or verifiable statement.

1. What is the probability that snakes could talk? (1 – 5)?
2. What is the probability that God could have made humans immortal by eating from the tree of life (1 – 5)?
3. What is the probability that humans achieved knowledge by eating fruit (1 – 5)?
4. The serpent used the plural of Gods as a level of status in talking to Eve. What is the probability that this was written independent of polytheistic religions? (1 – 5)?

5. What is the probability that we wear clothing because Adam and Eve became cognizant of their nakedness? (1 – 5)?
6. Adam and Eve did not die from eating fruit of the tree of knowledge. What is the probability that God changed the laws of physics? (1 – 5)?
7. Knowledge of good and evil made Adam and Eve cognizant of nudity and they had to cover themselves to prevent being seen by a super-power. The story tells us that the super power, God, had the ability to create life, put them to sleep and perform operations, etc. What is the probability that God is concerned about seeing man and woman naked? (1 – 5)?
8. God had trouble locating Adam and Eve after they consumed the fruit. What is the probability that God still had infinite powers? (1 – 5)?

Chapter 8: Leaving the Garden

Book of Genesis Chapter 3

3:14 And the LORD God said unto the serpent, Because thou hast done
this, thou art cursed above all cattle, and above every beast of the field,
upon thy belly shalt thou go, and dust shalt thou eat all the days of thy
life:
3:15 And I will put enmity between thee and the woman, and between
thy seed and her seed, it shall bruise thy head, and thou shalt bruise his
heel.
3:16 Unto the woman he said, I will greatly multiply thy sorrow and thy
conception, in sorrow thou shalt bring forth children, and thy desire
shall be to thy husband, and he shall rule over thee.
3:17 And unto Adam he said, Because thou hast hearkened unto the
voice of thy wife, and hast eaten of the tree, of which I commanded
thee, saying, Thou shalt not eat of it: cursed is the ground for thy sake,
in sorrow shalt thou eat of it all the days of thy life,
3:18 Thorns also and thistles shall it bring forth to thee, and thou shalt
eat the herb of the field,
3:19 In the sweat of thy face shalt thou eat bread, till thou return unto
the ground, for out of it wast thou taken: for dust thou art, and unto dust
shalt thou return.
3:20 And Adam called his wife's name Eve, because she was the mother
of all living.
3:21 Unto Adam also and to his wife did the LORD God make coats of
skins, and clothed them.
3:22 And the LORD God said, Behold, the man is become as one of us,
to know good and evil: and now, lest he put forth his hand, and take also
of the tree of life, and eat, and live for ever:
3:23 Therefore the LORD God sent him forth from the garden of Eden,
to till the ground from whence he was taken.
3:24 So he drove out the man, and he placed at the east of the garden of
Eden Cherubims, and a flaming sword which turned every way, to keep
the way of the tree of life.

God punishes the serpent but it is not said God took away his speech. God reduces the capital punishment to inconveniences for the human race forever. Generally, the authors state to whom God is addressing and if he changes the receiver. God was talking to the serpent in 3:14 and it appears he is continuing in 3:15 since the story does not change who God is directing commands. So, God creates a mutual dislike and mistrust between the serpent and the woman. The second part of the statement is a bit cryptic, rather than try to derive meaning beyond the apparent, deeper interpretations will be left for scholars. The simple intent is that future generations of snakes and mankind (humans) will hold animosity that humans will bruise snakes' heads, and snakes will bruise mans' heel. Here it seems the authors of Torah were not aware of the existence of poisonous snakes or large constrictors. These poisonous snakes or large constrictors tend to be a bit more hazardous to humans than a bruised heel.

God punished the woman and thus all following female humans by making child bearing dreary and childbirth painful. The writers of the Torah understood a woman's pain in childbirth but animals did not seem to show as much pain. Adding this curse in the story as God's command explained this apparent difference between human childbirth and other animals. Many billions of women have suffered through childbirth during the existence of the human race, so far. This was quite a curse for gaining knowledge! The second part of the curse on the woman and thus future generations of women was that their husbands shall govern their desires and they shall be subservient to their husbands. Setting the role of women as subservient to men was either God directing the writers to state this because the Bible is the inerrant word of God or the authors, redactors and scribes included the curse to all women to place them in a subservient roll from personal and religious beliefs at the time of writing and editing. The authors, redactors and scribes were very likely all male. This story was written and edited over a thousand years preceding the Bible being canonized. Gender conflicts are a very basic part of human life. Unfortunately, the original scriptures do not exist or have never been found, so we do not have a copy of the earliest writings. The earliest writings available are the Dead Sea Scrolls and the opening of Genesis has never been located in enough detail to piece together. So, the Bible is the source and the editing and scribing of the sources were done from 1000BCE to around 150CE. Cleopatra reigned Egypt from 69 to 30 BCE. Nefertiti was the wife of Egyptian Pharaoh Akenaten (Father of Pharaoh Tutankhamum). Akenaten and his wife Nefertiti rule Egypt from 1370 to 1330

BCE and she commanded great power. These two women demonstrated women were capable of ruling great empires and one ruled approximately at the early writing of the Bible and one rule close to the end of editing the Bible. There is no scientific evidence that these women influence the inclusion of the restriction of women but it seems likely. With no actual proof, we have to use the highest probability cause and it appears that gender conflicts inspired the constraints to be added of women's roles to the Bible more than divine intervention[7].

Now God turns his wrath to Adam for eating fruit of the forbidden tree. First God disallows eating fruit of the forbidden tree. Second God cursed the earth that it may not provide in abundance but produce thorns and thistles. Man has to cultivate the earth to make it produce and then man should only eat the herb of the field. God commands man to live a vegan lifestyle. God states that only by hard labor will man eat, and God reminds him, he is made of dust and will return to dust. Eating the fruit of knowledge and the punishment is similar to the mythologies Jews obtained during travels to Egypt and Mesopotamia and became tribal knowledge used by the authors during writing the Bible. Other religions had deities that became immortal. The condemnation of Adam to mortality and this statement condemns man to mortality to distance the Jewish faith from polytheistic religions of the time and explain man's mortality. The eating of the fruit and punishment will be re-emphasized later.

God makes clothing of skins. The skins have to be presumed to be animal. Two problems created by making animal skin clothes are that man was restricted to a vegetarian diet yet he is given animal skins for clothes and also, for whom did Adam and Eve have to dress. Textile mills require time to construct and man has to learn which plants to grow to produce fibers for making thread and eventually cloth. Primitive man, absence of divine guidance, would require animal skins for material, as cloth would require generations to produce. But Adam was given skins from the creator of the universe possessing infinite power, wisdom and creativity, this seems to be a contradiction in skills. God can create life, but has trouble creating cotton jeans and denim work shirts. If man is taught to kill for material to dress, then killing for food seems minor or at least a no worse sin. Why did God restrict man to eat only herbs of the field? No other people existed and God did not object to Adam and Eve being naked before they eat of the tree of knowledge of good and evil, so why did he insist on clothing them after. Here we want to give the benefit of doubt and say it was compassion for Adam and Eve being self-conscience and he was comforting they emotional states. But comforting Adam and Eve seems a bit contradictory after the curse of women to painful child birth for eternity, requiring man cultivate ground to grow crops and restricting them to a vegetarian diet. And even more punishment is coming so giving them clothes to reduce their inhibitions and uneasiness is not appropriate with the punishment.

The story reverts back to the mythology level of the time of writing. Both Egyptian and Mesopotamian (Assyrian and Babylonian) religions allowed mortals to become immortal through different means and even become equal to the Gods. Verses of Genesis 3:23 through 3:24 show strong evidence the writers of the bible were converting the polytheistic mythologies back to monotheistic[8]. God says "Behold, the man is become as one of us, to know good and evil". God says "one of us", plural! He is not talking to Adam or Eve, as this does not fit communication syntax. God has to be talking among other equal deities to use the pronoun us. And God says it with a bit of alarm since he interjects the explicative, behold. God is not alarming Adam and Eve that they are becoming God-like. God is alarmed that if they eat of the tree of life then Adam and Eve gain immortality and become Gods. We know this did not happen since we are all mortal and the writers had to rationalize human mortality. To ensure that humans remained mortal, Adam and we assume Eve were driven from the Garden of Eden. The make sure they did not reenter God placed Cherubim, and a flaming sword, which turned every way to keep them out. Cherubim are not defined in the Bible but were use in architecture by Greeks, Egyptians, Romans, and Assyrians generally as a animal body, of lion, horse or goat and human head and winged. It is believed that Cherubim were guards with status lower than angels. The reference to Cherubim occurs several times in the Bible without a good explanation. The flaming sword seems fairly descript. The two guards prevented Adam and Eve from returning.

Questions over Chapter 8

1) God disallows future snakes to talk (even though not listed in the bible)
   a) as a punishment for eating from the tree God placed in the Garden.
   b) to align snakes with today's species.
   c) because snakes are evil.
   d) because snakes bruise man's heel.
2) God punishes snake more by
   a) not allowing them to eat fruit.
   b) making them crawl in the dust.
   c) placing them in an adversarial roll with humans.
   d) making them crawl in the dust and placing them in an adversarial roll with humans.
3) God punishes all women
   a) because they were lower status than men.
   b) because Eve defied Him.
   c) by making child birth desirable.
   d) because Eve talked to the serpent.
4) The second part of women's punishment
   a) was to never talk to snakes.
   b) to not eat of the tree of knowledge again.
   c) be subservient to husbands.
   d) we are not told what the punishment was.
5) God had directed Adam not to eat of the tree of knowledge so
   a) Adam died as God had said.
   b) man must work the field to eat.
   c) man cannot talk to women.
   d) he could live forever.
6) Adam and Eve had to leave the Garden of Eden
   a) to avoid the snake.
   b) as punishment for wearing fig leafs.
   c) for hiding from God
   d) to prevent them from eating of the tree of life.

7) God placed Cherubim, and a flaming sword at the Garden of Eden
   a) to keep people out.
   b) to keep Adam and Eve from eating of the tree of life.
   c) for protection.
   d) to make it more exotic.
8) God said "Behold, the man is become as one of us" appears he
   a) was talking to Adam.
   b) he was talking to Eve.
   c) he was talking to other Gods.
   d) he considered himself as a human.

Probability of statements in Chapter 8

Instructions: Assign a number to the plausibility of each statement.
- (1) For a total contradiction to a statement.
- (2) For an implausible statement.
- (3) For a possible and maybe even practical statement.
- (4) For a plausible statement.
- (5) For a confirmed or verifiable statement.

1. Is it probable that all women have suffered through extreme discomfort of childbirth because of Eve's sin? (1 – 5)?
2. What is the probability that serpents once talked to communicate with humans? (1 – 5)?
3. What is the probability that women are lower status than men because of the original sin? (1 – 5)? (Note that women may be considered of lower status due to social, economic, political, and religious standards imposed by society, and bad TV journalism but the question imposed by the Biblical story is independent of these values, and is asking is a female life of less intrinsic value than a male life.)
4. What is the probability that writers of the Bible were not threatened by women assuming power and only repeated God's direction to keep women as lower status than men? (1 – 5)?
5. What is the probability that eating of the tree of life would have made humans immortal? (1 – 5)?

6. God seems to discuss the original sin and the prevention of humans becoming immortal with other Gods. What is the probability that God would include this statement in the inerrant word of God? (1 – 5)?
7. Cherubim and flaming swords guarding the Garden of Eden make it stand out from unguarded gardens. What is the probability that Cherubim and flaming swords actually exist to protect the Garden of Eden? (1 – 5)?

Chapter 9: Starting the first family

Book of Genesis Chapter 4

4:1 And Adam knew Eve his wife, and she conceived, and bare Cain, and said, I have gotten a man from the LORD.
4:2 And she again bare his brother Abel. And Abel was a keeper of sheep, but Cain was a tiller of the ground.
4:3 And in process of time it came to pass, that Cain brought of the fruit of the ground an offering unto the LORD.
4:4 And Abel, he also brought of the firstlings of his flock and of the fat thereof. And the LORD had respect unto Abel and to his offering:
4:5 But unto Cain and to his offering he had not respect. And Cain was very wroth, and his countenance fell.
4:6 And the LORD said unto Cain, Why art thou wroth? and why is thy countenance fallen?
4:7 If thou doest well, shalt thou not be accepted? and if thou doest not well, sin lieth at the door. And unto thee shall be his desire, and thou shalt rule over him.
4:8 And Cain talked with Abel his brother: and it came to pass, when they were in the field, that Cain rose up against Abel his brother, and slew him.
4:9 And the LORD said unto Cain, Where is Abel thy brother? And he said, I know not: Am I my brother's keeper?

Adam and Eve had two sons. Now the world's population has doubled to four. Cain the first-born became a farmer and raised crops and his later brother Abel became a rancher and raised sheep. The commandment to sacrifice animals and crops is mentioned more than any other commandment in the Bible. This is the first occurrence and it has unexplained connotations with it. In the chapters explaining many of the laws and commandments, the rest of Genesis, Exodus, Leviticus, Numbers, etc. great detail is given to the method of sacrificing animals and several times it is mentioned that the first part of crops must be given in sacrifice. God is pleased with the first lambs of the flock that Abel offered in sacrifice but God is not pleased with the fruit or harvest of the ground. God does not know that the sheep had to eat plants grown from the ground to get fat and healthy. But yet God distinguishes between herding sheep and raising crops. There is no indication as to the reason for this distinction. Cain is distraught that God prefers Abel's sacrifice to his. God offers no comfort to Cain in questioning him. Cain kills Abel and now the world population is reduced to three. God questions Cain as to his brother's location and Cain lies to God. God that has created the world and everything in it is not cognizant of a murder of one fourth of the population.

Book of Genesis Chapter 4

4:10 And he said, What hast thou done? the voice of thy brother's blood crieth unto me from the ground.
4:11 And now art thou cursed from the earth, which hath opened her mouth to receive thy brother's blood from thy hand,
4:12 When thou tillest the ground, it shall not henceforth yield unto thee her strength, a fugitive and a vagabond shalt thou be in the earth.
4:13 And Cain said unto the LORD, My punishment is greater than I can bear.
4:14 Behold, thou hast driven me out this day from the face of the earth, and from thy face shall I be hid, and I shall be a fugitive and a vagabond in the earth, and it shall come to pass, that every one that findeth me shall slay me.
4:15 And the LORD said unto him, Therefore whosoever slayeth Cain, vengeance shall be taken on him sevenfold. And the LORD set a mark upon Cain, lest any finding him should kill him.
4:16 And Cain went out from the presence of the LORD, and dwelt in the land of Nod, on the east of Eden.

Murder did not escape God's notice, so God's questioning Cain was to torment him. The lack of understanding of nature and biological science is revealed in the mystical reference of the earth swallowing Abel's blood. God informs Cain that the earth will no longer yield the best crops and he will struggle and wander to survive as a fugitive. Cain tells God his punishment is more than he can bear. Cain states that driven from the face of the earth that he has known and hiding from God will leave him vulnerable to everyone he meets that they may kill him. At this time there are three people on earth, Cain, Adam and Eve. Maybe he fears his mother and father will kill him but this seems unlikely. Who are these other people Cain fears? To ease Cain's fears God marks him such that if anyone kills Cain their death shall be seven times worse. So Cain leaves the land he has known and goes east of Eden to Nod. At the time of composing the Bible, the practice was to name a land after the ruler, so who named the new land Nod?

Book of Genesis Chapter 4

4:17 And Cain knew his wife, and she conceived, and bare Enoch: and he builded a city, and called the name of the city, after the name of his son, Enoch.
4:18 And unto Enoch was born Irad: and Irad begat Mehujael: and Mehujael begat Methusael: and Methusael begat Lamech.
4:19 And Lamech took unto him two wives: the name of the one was Adah, and the name of the other Zillah.
4:20 And Adah bare Jabal: he was the father of such as dwell in tents, and of such as have cattle.
4:21 And his brother's name was Jubal: he was the father of all such as handle the harp and organ.
4:22 And Zillah, she also bare Tubalcain, an instructer of every artificer in brass and iron: and the sister of Tubalcain was Naamah.
4:23 And Lamech said unto his wives, Adah and Zillah, Hear my voice, ye wives of Lamech, hearken unto my speech: for I have slain a man to my wounding, and a young man to my hurt.
4:24 If Cain shall be avenged sevenfold, truly Lamech seventy and sevenfold.
4:25 And Adam knew his wife again, and she bare a son, and called his name Seth: For God, said she, hath appointed me another seed instead of Abel, whom Cain slew.
4:26 And to Seth, to him also there was born a son, and he called his name Enos: then began men to call upon the name of the LORD.

(Heritage descending from Cain is delineated in Chapter 10 for better understanding)

Cain finds a wife! It is not explained where Cain found a wife when only three human beings exist on earth. Cain and his wife have a son, Enoch. Cain builds a city and names it Enoch. The first city (which, by the way in this part of the world was Jerusalem not Enoch) must have had not utilities or services. Water distribution and sewage removal were unknown. Electricity was not discovered so there were no power or communication distributions. Still building a city was a great accomplishment for one man. Who was going to live in this city of Enoch? Adam and Eve[9], Cain and his unnamed wife and their son Enoch are the only residences of this area. Now several generations are started from Enoch. How did Enoch find a wife? There are no other people! Enoch found a wife. The next generation Enoch's son, Irad found a wife. The next generation Irad's son, Mehujael found a wife. Finding wives from non-existent population continues for five generation. At some point in the heritage there had to be some daughters born and not written into the story. These daughters would have had the same family genetics and would started the human race from inbreeding if they became wives to the sons.

The fifth generation descending from Cain was Lamech and he took two wives. The wife Adah had two sons, one son Jabal, initiated the line of nomadic herders and the other son Jubal, initiated the line of ancient musicians. The other wife had a son and a daughter. The son Tubalcain instructed use the use of brass and iron for creating artifacts. The Bronze Age started in northern Europe around 1800BCE and the Iron Age developed around 600 to 500 BCE. Iron work in the Roman Empire started a few hundred years later and was well distributed through the area where the Torah and books for the Bible were being edited during its final edits and canonizing. Brass, Bronze and iron were not being used at the timeframe of the story creation. So when Tubalcain was instructing future generations in the use of brass and iron it took several hundred generations for its use to become wide spread. Tubalcain's sister Naamah seems to have little role in developing the culture or at least the male redactors, scribes and writers seemed to over look this. The full family story may be included for the sake of Lamech's tragedy. This including of his full family of two wives, and four children is to demonstrate a full life and then he has to announce his dreadful deed. Lamech says he has killed a man and a young man. It is not clear if he has killed a man and also a young man or if he has killed one man and to stress the heinousness of the crime, he emphasizes that he was a young man. The verse is also not clear if Lamech is trying to stress that he has killed because he was wounded and hurt and thus committed the crime in defense or if the sin of killing is such a burden and bothering his conscience that the act of killing has wounded him and hurt him. But Lamech is able to recall five generations past that Cain had killed and Lamech relates his sin as eleven times worse.

The story line changes back to Adam and Eve. Eve produces another son, Seth. If the chronological order is maintained, this is after eight generations of human beings have been born or the seventh generation succeeding Adam and Eve. Seth finds a wife or concubine to bear a son, Enos. Enos is significant in returning people to the Lord or God. This line of descendants will increase in importance as the story progresses.

Questions over Chapter 9

1. Adam and Eve have two sons (ignoring Seth for now)
    a. Cain first born the farmer and Abel the sheep rancher.
    b. no daughters that we are told.
    c. Cain the older kills Abel the younger.
    d. all the above.
2. The sacrifices to God were
    a. accepted and pleased God.
    b. to show the first families appreciation to God.
    c. fattened lambs from Cain and part of the grain harvest from Abel.
    d. all the above.
3. God accepted the sacrifice graciously
    a. that Cain and Abel had made.
    b. equally well for animals and grains.
    c. of the animals but not the grain.
    d. of Cain but not Abel.
4. Cain kills Abel
    a. because he did not like sheep herders.
    b. because Adam and Eve liked Abel best.
    c. to show killing is wrong.
    d. because he was distraught over God's preference in sacrifices.
5. God approaches Cain about killing Abel
    a. because He cannot find Abel.

   b. we are not told why but it appears to be part of a punishment.
   c. because God cannot conceive why Cain would agonize over his sacrifice being eschewed.
   d. all the above.
6. God tells Cain he must leave and live in Nod east of Eden and Cain
   a. fears for his life from the people who will kill him because he killed Abel.
   b. will miss the farm he started.
   c. does not want to leave Adam and Eve.
   d. is not sure where Nod is because the maps were not made yet.
7. Jabal and Jubal, sons of Lamech and Adah were
   a. herdsmen.
   b. farmers.
   c. a herdsman and farmer.
   d. a herdsman and music instructor.
8. Tubalcain instructed
   a. religion to please God.
   b. use the use of brass and iron for creating artifacts.
   c. proper sacrificial practices to avoid the Cain Abel catastrophe.
   d. the initiation of the line of ancient musicians.
9. Lamech kills a young man
   a. in self-defense so it is no problem.
   b. because he was young.
   c. for money.
   d. which makes his burden 11 times worse than Cain's killing Abel.
10. Adam and Eve have another son
   a. Seth
   b. Enos
   c. Abel the second.
   d. Enoch.

Probability of statements in Chapter 9

Instructions: Assign a number to the plausibility of each statement.

(1) For a total contradiction to a statement. (2) For an implausible statement. (3) For a possible and maybe even practical statement. (4) For a plausible statement. (5) For a confirmed or verifiable statement.

1. Cain was able to find a wife when only 3 people exist. What is the probability that both of these are correct? (1 – 5)?
2. Cain is afraid of leaving the land he has known because people will kill him. But at this time only 3 people exist. What is the probability that both of these are correct? (1 – 5)?
3. Cain and his wife of unknown origins have a son Enoch. Enoch gets married and has a son. What is the probability that there was a woman for Enoch to marry? (1 – 5)?
4. Enoch's son, Irad found a wife What is the probability that there was a woman for Irad to marry? (1 – 5)?
5. The fifth generation, Tubalcain instructed use the use of brass and iron for creating artifacts. This places the creation 5 generations before the middle of the Roman Empire's existence, which would at best make the creation in the early part of the Roman Empire's beginning. What is the probability that the writers of the Bible set a correct time line? (1 – 5)?
6. What is the probability that Lamech can recall his relative's sin from five generations past and calculate it as eleven times worse? (1 – 5)?
7. Jerusalem according to excavations is the oldest city in the Middle-East, not Enoch, but the third person on earth takes a wife and has a son, Enoch that built Enoch. What is the probability that Enoch is older than Jerusalem and ruins have never been found? (1 – 5)?
8. If all people started from Cain, descending from Adam and Eve, the daughters and sons had to inbreed to propagate the human race. What is the probability that this is correct? (1 – 5)?

9. Jubal instructed a line of musicians that handle harp and organs. After only five generations have existed on earth what is the feasibility of this? (1 – 5)?

Chapter 10: Creation summary and humans beings start

Book of Genesis Chapter 5

5:1 This is the book of the generations of Adam. In the day that God created man, in the likeness of God made he him,
5:2 Male and female created he them, and blessed them, and called their name Adam, in the day when they were created.
5:3 And Adam lived an hundred and thirty years, and begat a son in his own likeness, and after his image, and called his name Seth:
5:4 And the days of Adam after he had begotten Seth were eight hundred years: and he begat sons and daughters:
5:5 And all the days that Adam lived were nine hundred and thirty years: and he died.
5:6 And Seth lived an hundred and five years, and begat Enos:
5:7 And Seth lived after he begat Enos eight hundred and seven years, and begat sons and daughters:
5:8 And all the days of Seth were nine hundred and twelve years: and he died.
5:9 And Enos lived ninety years, and begat Cainan:
5:10 And Enos lived after he begat Cainan eight hundred and fifteen years, and begat sons and daughters:
5:11 And all the days of Enos were nine hundred and five years: and he died.
5:12 And Cainan lived seventy years and begat Mahalaleel:
5:13 And Cainan lived after he begat Mahalaleel eight hundred and forty years, and begat sons and daughters:
5:14 And all the days of Cainan were nine hundred and ten years: and he died.
5:15 And Mahalaleel lived sixty and five years, and begat Jared:
5:16 And Mahalaleel lived after he begat Jared eight hundred and thirty years, and begat sons and daughters:
5:17 And all the days of Mahalaleel were eight hundred ninety and five years: and he died.
5:18 And Jared lived an hundred sixty and two years, and he begat Enoch:
5:19 And Jared lived after he begat Enoch eight hundred years, and begat sons and daughters:
5:20 And all the days of Jared were nine hundred sixty and two years: and he died.
5:21 And Enoch lived sixty and five years, and begat Methuselah:
5:22 And Enoch walked with God after he begat Methuselah three hundred years, and begat sons and daughters:

5:23 And all the days of Enoch were three hundred sixty and five years:

Book of Genesis Chapter 5

5:24 And Enoch walked with God: and he was not, for God took him.
5:25 And Methuselah lived an hundred eighty and seven years, and
begat Lamech.
5:26 And Methuselah lived after he begat Lamech seven hundred eighty
and two years, and begat sons and daughters:
5:27 And all the days of Methuselah were nine hundred sixty and nine
years: and he died.
5:28 And Lamech lived an hundred eighty and two years, and begat a
son:
5:29 And he called his name Noah, saying, This same shall comfort us
concerning our work and toil of our hands, because of the ground which
the LORD hath cursed.
5:30 And Lamech lived after he begat Noah five hundred ninety and
five years, and begat sons and daughters:
5:31 And all the days of Lamech were seven hundred seventy and seven
years: and he died.
5:32 And Noah was five hundred years old: and Noah begat Shem,
Ham, and Japheth.keeper?

Chapter 4 of Genesis covers the first two sons of Adam well but they are not continued into the development of future generations. The descendants of Cain will be eliminated in the great flood. Naming the wives of the male heirs was not significant except for Lamech having two separate wives. Adam and Eve have another son Seth when Adam is 130 years old. The following diagram of the heritage descending from Adam and Eve is to simplify the text of Chapter 5 of Genesis.

God creates:
↳**Adam and Eve** Adam and Eve had a third son Seth when Adam is 130 and Adam lives 930 years
  ↳Cain and Able; Able dies before baring children
    ↳Enoch
      ↳ Irad
        ↳Mehusael
          ↳ Lamech with 2, Adah and Zillah
            ↳Jabal, Jubal, Tabalcain and Naamah

  ↳Seth had Enos at 105 and lived 912 years, had many unknown sons and daughters
    ↳Enos had Cainan at 90 and lived 905 years, had many unknown sons and daughters
      ↳Cainan had Mahalaleel at 70 and lived 910 years, had many unknown sons and daughters
        ↳ Mahalaleel had Jared at 65 and lived 905 years, had many unknown sons and daughters
          ↳ Jared had Enoch at 162 and lived 962 years, had many unknown sons and daughters
            ↳ Enoch had Methuselah at 300 and lived 969 years, had many unknown sons and daughters
              ↳ Methuselah had Lamech at 187 and lived 969 years, had many unknown sons and daughters
                ↳Lamech had Noah at 595 and lived 770 years, had many unknown sons and daughters
                  ↳Noah It is hinted that Noah had 3 sons at 500 years old
                    ↳ Shem, Ham and Japheth

The many children were unnamed since they would perish in the great flood. Wives remained unnamed to continue the trend of minimizing women's role in society. The names of descendants are confused between Seth and Cain. Cain had Enoch and so did Jared. Enoch descending from Seth had Methusael and very similar with slight change in spelling is Methuselah and Lamech is a descendant of both Methusael and Methuselah. The population of the world is still quite small at this time and perhaps the development of names is primitive at best, but with such a small population it is very unlikely that names would be duplicated. Is the heritage being out of order in the two chapters of Genesis an error in writing the history or editing the or are these actual duplications in names? The ages also are not possible under the same physical laws that apply today. Did God change the physical laws or change the length of a year. If a year were one-tenth as long these may be possible and stretching the limits of feasibility. It is misleading for the inerrant word of God to give us these apparently erroneous facts with no explanation.

Questions over Chapter 10

1. Why was there such detail on Cain and Abel if they are not the start of the Human race?
    a. We are not told.
    b. To provide moral guidance.
    c. Because man should not be alone.
    d. To make the story complete.
2. Why did Adam and Eve have Seth?
    a. We are not told.
    b. Cain was a killer and could not be the start of the human race.
    c. Able was dead.
    d. Adam and Eve were only 130 years old and time for a second generation.
3. Why were names repeated from Cain's descendants?
    a. The population was small so there were not a large number of names.
    b. It appears the scribes and redactors were careless in this chapter.

   c. The names were the same for different people.
   d. It is hard to read and the descent line cannot be distinguished.
4. In this story, life expectancy was close to 1000 years
   a. and children were conceived as late as 500-year-old.
   b. because the population was small.
   c. food and health were better.
   d. because God like people better back then.
5. Only male descendants were listed
   a. because there were no daughters.
   b. to continue the trend in the Bible of minimizing women's roles.
   c. because daughters were not named.
   d. to minimize writing as it was very difficult.
6. If all people started from Adam and Eve
   a. men had to have children with their very close relatives.
   b. women were found other than what God created.
   c. genetics could recover after thousands of generations.
   d. all the above.
7. How was this genealogy constructed?
   a. It could only be the word of God.
   b. Writers of the Bible traced county records.
   c. We are not told, but it seems most likely to be writer's fantasy.
   d. The family was so limited it was easy to trace.

Probability of statements in Chapter 10

Instructions: Assign a number to the plausibility of each statement.

(1) For a total contradiction to a statement.
(2) For an implausible statement.
(3) For a possible and maybe even practical statement.
(4) For a plausible statement.
(5) For a confirmed or verifiable statement.

1. What is the probability that the writers did not mix up names and they were reused (1 – 5)?
2. What is the probability that people lived close to 1000years (1 – 5)?
3. What is the probability that God change the physical laws to allow longer life (1 – 5)?
4. What is the probability that Lamech having two separate wives support Adams declaration of marriage of a man cleaving unto his wife (1 – 5)?
5. What is the probability that men found spouses only within the descendants of Adam and Eve (1 – 5)?
6. What is the probability that God intended women to have inferior roles in life so He wrote the Bible listing only male descendants with a few exceptions (1 – 5)?
7. What is the probability that the entire population started from Adam and Eve (1 – 5)?
8. What is the probability that record keeping at the beginning of the human race was adequate to produce the family tree over 10 generations after Adam and Eve (1 – 5)?

Chapter 11: The World develops in Sin

Book of Genesis Chapter 6

6:1 And it came to pass, when men began to multiply on the face of the earth, and daughters were born unto them,
6:2 That the sons of God saw the daughters of men that they were fair, and they took them wives of all which they chose.
6:3 And the LORD said, My spirit shall not always strive with man, for that he also is flesh: yet his days shall be an hundred and twenty years.
6:4 There were giants in the earth in those days, and also after that, when the sons of God came in unto the daughters of men, and they bare children to them, the same became mighty men which were of old, men of renown.

It is very difficult to be objective and not to be judgmental in reading chapter 6 of Genesis. The objective of this textbook is the science of creation as stated in the Bible so a critical approach is required to ensure the integrity of the source. God observes the world as the population increases. Humans are apparently following instincts that are present in modern day human beings, that is, men find women attractive and they followed the same instincts that people today follow. There is no statement contrary to humans having the same instincts towards relationships and the only laws of marriage have been given as Adam took Eve that they are to cleave unto one another. The people descending from Adam and Eve are having children and multiplying so the population grows. To this time God has made one commandment to not eat of the tree of knowledge. God has banished humans from the Garden of Eden and placed a flaming sword and Cherubim to guard reentry by the humans. So, eating of the tree of knowledge is not a problem. The other direction given was that man was to eat of the herbs of the field or be strict vegetarian. God gave an indication, although it was not a direct command that he preferred Abel's sacrifice of animals to Cain's sacrifice of crop or herbs of the field. Only Cain knew this preference shown by God and there is no evidence that Cain distributed this fact. So we are not told if people are consuming animals for food and even if they are, this would be contradictory for

God to punish them for eating meat when He clearly preferred animal sacrifices to sacrifices of crops. God clothed Adam and Eve with animal skins instead of creating a loom to make cloth from plant fibers.

Note that God distinguishes between the sons of God and daughters of men. The writers did not specify from where these daughters of men were coming. This verse does not denote that this is an unacceptable arrangement with God or that it was a sin. But it is clearly mentioned and seems that to mention it would have some purpose. Were the men that conceived the daughters that were designated as the daughters of men, not the sons of God? If not, from where did these men come?

There are eight generations leading to Noah that could multiply to create the population[10]. If every woman of each generation had the maximum number of children and assuming there are equal numbers of men and women and they are matched in pairs, then the maximum population of the world be 1,100,000. This would follow from seven reproductive cycles where ten boys and ten girls are born to each woman and they are not polygamist. This number 1,100,000 also assumes the descendants of Cain are separate and unique from Seth's even with the same name and that Cain's descendants had the maximum number of children. Present day studies show that with primitive food supply and health care, a growth rate of this order is not feasible. God condemned man to work hard to force the earth to yield minimum herbs of the field and not provide the abundance that was in the Garden of Eden, so food and health care were primitive. Realistically a ten percent growth rate of this population would be an extremely high estimate. Nine generations of ten percent growth rate even rounding up for benefit of doubt creates less than 100 people on earth.

God points out that humans are mortal because His spirit is not with them and they shall be limited to 120 years. God points out that his spirit will not always be with man and that they are of flesh. This emphasizes that humans are mortal, but this also hints at the fact that humans have a will independent of God's. If God's spirit is not with humans, then they have to use their own judgment. This is an important point in the following books of Genesis. Chapter 5 of Genesis lists eight generations that lived an average of 848 years with the shortest life span being 365 and longest life span near 900. Maybe God did not read Genesis chapter 5 and if He wrote it, He must have forgotten.

The earth having giants is a bit obscure. Were some men or women very large as present day pro basketball stars or does this refer to unnaturally large as mythical characters, say Paul Bunyan type. In either case does being this large constitute a sin? God created all creatures in previous chapters and this verse would indicate the presence of beings He did not create. This maybe a sore spot with God if giants exist that He did not create. The presence of giants seems to be part of the aggravation to God making Him distraught with the world. The rest of the statement hints at the displeasure God is finding. When sons from the descendants God started, had children with daughters not listed, the children could become strong and self-sufficient, which seems to be an aggravation with God. We are not told why this is an aggravation. Where are these daughters that were not children of God coming from? A consistent thread through the story of the creation is the extra people not created by God keep popping into the story. Why are these people so prevalent? Genesis 6 so far has explained that God is upset about the presence of giants and men becoming self-sufficient by marrying women that were not of children of His creation.

Book of Genesis Chapter 6

6:5 And God saw that the wickedness of man was great in the earth, and that every imagination of the thoughts of his heart was only evil continually.
6:6 And it repented the LORD that he had made man on the earth, and it grieved him at his heart.
6:7 And the LORD said, I will destroy man whom I have created from the face of the earth, both man, and beast, and the creeping thing, and the fowls of the air, for it repenteth me that I have made them.

God saw the wickedness of man. We cannot argue with God, but science demands we be inquisitive and question why is God upset and this is the science of creation, not the religion. The only instructions given by God were listed above as do not eat of the tree of knowledge, be a strict vegetarian, and God prefers animal sacrifices to grains or vegetable. Just as a note of humor, it is duplicitous for God to command a vegetarian diet and prefer animal sacrifices. Through eight generations man has to establish codes of conduct for a small society to survive. These would have to be fairly basic and simple to guide people to coexisting and surviving. The inference to wickedness and that the imagination of the thoughts were only continually evil suggest that God was judging people on a more sophisticated level than basic survival. The evolution of laws cannot be discussed, as it would entail the development of the human race for several hundred thousand years leading to the writing of the Bible. The writers returning from Babylon had access to knowledge of the Hammurabi's law code from 3750 years ago. Hammurabi's law code was written on cuneiform and is still preserved in the Louver Museum in Paris. Hammurabi's law would have existed at the beginning of the writing of the Bible. The use of God's judgment of people based on more sophisticated laws than the people were given suggests the writers were evaluating the existing circumstance during the time of writing the Bible rather than during the era for which they were describing. The criticalness of God's judgment appears to be aligned with Hammurabi's law code or laws in place at the time of writing the Bible, more than with the basic rules He has given in the story.

So, God has ill feelings about starting the human race. Why? How does man's wickedness and that the imagination of his thoughts were only continually evil, hurt God? Even deeper is how could the thoughts be evil and wicked when they are following the only few simple commands they were given? There is no answer to this and religious leaders tell us, we are not allowed to question God's actions or thoughts[11].

So, to continue let's allow God to be upset. God decides to destroy man and all creatures on earth or that fly. God does not decide to destroy all sea creatures. Why does God wish to destroy all animals? Are the birds and animals adding to man's wickedness and evil thoughts? Today scientists have extensively studied many animals and found social behavior that has developed over time that assist in the survival of the species, but this would have not been known at the time of writing the Bible. So, the destruction of all life on earth seems to be a carryover from the Egyptian and Assyrian religions collected by the Jews during exile to these lands.

Tablet 11, the Flood Tablet from the Atrahasis Epic into the Gilgamesh Epic was written before the Biblical story and appears to be the main story the Bible's writers plagiarized. This is known as the Assyrian Flood Myth and shows in many web searches and Wikipedia but not for discussion in this text. The context of God's anger and desire to destroy the entire world with little explanation seems to be a common knowledge or tribal knowledge that does not require an explanation in writing this part of the Bible. In simple terms God sees people as evil and must destroy them. The only complete destruction the writers could conceive was a massive flood. Therefore, everything must be destroyed. Again, the trends of the story imply plagiarizing previous myths and religions and disregarding logic or analysis. The simplicity indicates a human level of writing over a treatise of divine origins. This will be analyzed in the probability of the statements at the end of the chapter.

Book of Genesis Chapter 6

6:8 But Noah found grace in the eyes of the LORD.
6:9 These are the generations of Noah: Noah was a just man and perfect in his generations, and Noah walked with God.
6:10 And Noah begat three sons, Shem, Ham, and Japheth.
6:11 The earth also was corrupt before God, and the earth was filled with violence.
6:12 And God looked upon the earth, and, behold, it was corrupt, for all flesh had corrupted his way upon the earth.
6:13 And God said unto Noah, The end of all flesh is come before me, for the earth is filled with violence through them, and, behold, I will destroy them with the earth.
6:14 Make thee an ark of gopher wood, rooms shalt thou make in the ark, and shalt pitch it within and without with pitch.
6:15 And this is the fashion which thou shalt make it of: The length of the ark shall be three hundred cubits, the breadth of it fifty cubits, and the height of it thirty cubits.
6:16 A window shalt thou make to the ark, and in a cubit shalt thou finish it above, and the door of the ark shalt thou set in the side thereof, with lower, second, and third stories shalt thou make it.
6:17 And, behold, I, even I, do bring a flood of waters upon the earth, to destroy all flesh, wherein is the breath of life, from under heaven, and every thing that is in the earth shall die.

God decides Noah is worth saving and his family will be spared as well. God tells Noah the world is corrupt and violent and He will destroy it. God tells Noah to build an ark, a water vessel for an extended floatation. God directs Noah to build the ark 300cubits[12] by 50 cubits wide by 30 cubits high. The ark in metric would be approximately 137m long by 23m wide by 14m tall. A boat this size would have been impressive for the time but to hold a pair of every species, this would have been a bit cozy. To help some this was divided into three stories or floors. With the thickness of the floors, this would make each deck with about a 4-meter ceiling if they were divided equally. There possibly could have been one floor with a higher ceiling for the taller animals, and two lower ceilings for all others. Then God tells Noah of his plan to destroy all life by a flood. Then God tells Noah to gather his wife, the three sons with their wives, and a male and female of every living thing on earth and to gather all the food they will need.

Book of Genesis Chapter 6

6:18 But with thee will I establish my covenant, and thou shalt come into the ark, thou, and thy sons, and thy wife, and thy sons' wives with thee.
6:19 And of every living thing of all flesh, two of every sort shalt thou bring into the ark, to keep them alive with thee, they shall be male and female.
6:20 Of fowls after their kind, and of cattle after their kind, of every creeping thing of the earth after his kind, two of every sort shall come unto thee, to keep them alive.
6:21 And take thou unto thee of all food that is eaten, and thou shalt gather it to thee, and it shall be for food for thee, and for them.
6:22 Thus did Noah, according to all that God commanded him, so did he.

For a two-day flood this maybe a large task, but God did not specify the period for which Noah should plan. Hay and grain may not be a storage problem, but carnivores require meat and storage of meat requires drying and processing to keep an extended period. Let's assume that the rain will last until the flood is over so rainwater may be caught to survive. This is an impossible task under the given conditions but let's ignore the problems and continue. The question remains why God would have Noah collect all animals when God made them in a day. God could destroy all animals and recreate them after the flood kills all life on earth. Noah did not question God and followed his instructions.

Book of Genesis Chapter 7

7:1 And the LORD said unto Noah, Come thou and all thy house into the ark, for thee have I seen righteous before me in this generation.
7:2 Of every clean beast thou shalt take to thee by sevens, the male and his female: and of beasts that are not clean by two, the male and his female.
7:3 Of fowls also of the air by sevens, the male and the female, to keep seed alive upon the face of all the earth.
7:4 For yet seven days, and I will cause it to rain upon the earth forty days and forty nights, and every living substance that I have made will I destroy from off the face of the earth.
7:5 And Noah did according unto all that the LORD commanded him.
7:6 And Noah was six hundred years old when the flood of waters was upon the earth.
7:7 And Noah went in, and his sons, and his wife, and his sons' wives with him, into the ark, because of the waters of the flood.
7:8 Of clean beasts, and of beasts that are not clean, and of fowls, and of every thing that creepeth upon the earth,
7:9 There went in two and two unto Noah into the ark, the male and the female, as God had commanded Noah.
7:10 And it came to pass after seven days, that the waters of the flood were upon the earth.
7:11 In the six hundredth year of Noah's life, in the second month, the seventeenth day of the month, the same day were all the fountains of the great deep broken up, and the windows of heaven were opened.
7:12 And the rain was upon the earth forty days and forty nights.
7:13 In the selfsame day entered Noah, and Shem, and Ham, and Japheth, the sons of Noah, and Noah's wife, and the three wives of his sons with them, into the ark,
7:14 They, and every beast after his kind, and all the cattle after their kind, and every creeping thing that creepeth upon the earth after his kind, and every fowl after his kind, every bird of every sort.
7:15 And they went in unto Noah into the ark, two and two of all flesh, wherein is the breath of life.
7:16 And they that went in, went in male and female of all flesh, as God had commanded him: and the LORD shut him in.
7:17 And the flood was forty days upon the earth, and the waters increased, and bare up the ark, and it was lift up above the earth.

Chapter 7 starts with a repeat of chapter 6 for those who cannot recall what they just read. Chapter 7 does add some detail not included in 6. First for clean beast Noah was to take 7, the male his female. This is not clear in today's English. This could be interrupted as 7 pairs of males and females or one male and six mates. If the small number is assumed of 7 total of each species, the ark is going to be quite full. Noah was directed for the non-clean beasts take only a pair of male and female. The writers of the Torah canonized into the Bible understood the difference between clean and unclean animals such as cattle and pigs so they felt no need to explain the difference here. The subject of clean and unclean animals is covered in more detail in Numbers, Leviticus, and Deuteronomy but lets consider it insignificant for the flood story as the number of animals already exceeds the limits of feasibility. Noah and his wife, their three sons and their wives and all the animals were onboard the ark and the Lord shut them in and the flood began. The flood covered the entire earth and everything died accept in the ark. The water remained 150 days. Why was this not the forty days that is commonly expressed in stories about the flood?

Book of Genesis Chapter 7

7:18 And the waters prevailed, and were increased greatly upon the earth, and the ark went upon the face of the waters.
7:19 And the waters prevailed exceedingly upon the earth, and all the high hills, that were under the whole heaven, were covered.
7:20 Fifteen cubits upward did the waters prevail, and the mountains were covered.
7:21 And all flesh died that moved upon the earth, both of fowl, and of cattle, and of beast, and of every creeping thing that creepeth upon the earth, and every man:
7:22 All in whose nostrils was the breath of life, of all that was in the dry land, died.
7:23 And every living substance was destroyed which was upon the face of the ground, both man, and cattle, and the creeping things, and the fowl of the heaven, and they were destroyed from the earth: and Noah only remained alive, and they that were with him in the ark.
7:24 And the waters prevailed upon the earth an hundred and fifty days.

Questions over Chapter 11

1) The population grows or multiplies to cover the face of the earth
   a) with daughters born of men.
   b) sons of God taking wives of daughters of men.
   c) because all daughters were chosen.
   d) all the above.
2) There were giants in those days
   a) which fought with sons of God.
   b) which God had to defeat to save his people.
   c) whose creation we were not told about.
   d) which explains many legends of giants.
3) Now God decides to limit life to
   a) 900 years.
   b) 120 years.
   c) 300 years.
   d) 45 years.
4) Humans are of flesh which
   a) is a reason for mortality.
   b) hints at having a free will.
   c) means God will not be with them.
   d) all the above.
5) God decided to destroy the earth because
   a) God saw the wickedness of man.
   b) God thought the wickedness of man was great in the earth.
   c) God thought that every imagination of the thoughts of man's heart was only evil continually.
   d) God repented that he had made man on the earth, and it grieved him at his heart.
   e) all the above.
6) God decided to save Noah and his family because
   a) Noah found grace in the eyes of the LORD.
   b) Noah was a just man and perfect in his generations.
   c) Noah walked with God.
   d) All the above.

7) Noah has to save the animal population by
   a) taking 7 of every clean animal and a pair of not clean animals.
   b) taking a pair of every species.
   c) getting one of every bird and animal.
   d) all the above.
8) The Flood lasted
   a) 40 days and 40 nights.
   b) as long as there were animals and people alive outside the ark.
   c) 150 days.
   d) one year.

Probability of statements in Chapter 11

Instructions: Assign a number to the plausibility of each statement.
(1) For a total contradiction to a statement.
(2) For an implausible statement.
(3) For a possible and maybe even practical statement.
(4) For a plausible statement.
(5) For a confirmed or verifiable statement.

1. What is the probability that God had reasons to be upset with the human race? (1 – 5)?
2. What is the probability that the human race could have grown to significant numbers in eight generations? (1 – 5)?
3. What is the probability that giants existed that were a race that God did not create? (1 – 5)?
4. What is the probability that God did not destroy all animals because He could not recreate them? (1 – 5)?
5. What is the probability that enough animals fit in the ark to repopulate the earth? (1 – 5)?
6. What is the probability that the ark could carry enough food and water for all species to live? (1 – 5)?
7. What is the probability that the writers of the Bible were relaying the story from the Jewish religion and not a fantasy of their own times and other religions and mythology? (1 – 5)?

Chapter 12: New beginning of the world

Book of Genesis Chapter 8

8:1 And God remembered Noah, and every living thing, and all the cattle that was with him in the ark: and God made a wind to pass over the earth, and the waters asswaged,
8:2 The fountains also of the deep and the windows of heaven were stopped, and the rain from heaven was restrained,
8:3 And the waters returned from off the earth continually: and after the end of the hundred and fifty days the waters were abated.
8:4 And the ark rested in the seventh month, on the seventeenth day of the month, upon the mountains of Ararat.

It rained for forty days but the water covered the earth for 150 plus days. Food and especially water had to be a problem. This seems to be a very large logical problem not related in the story. All civilization is gone and the world is starting new as the ark lands on mountains in Ararat. Ararat is as elusive as the Garden of Eden. The location of Ararat is not known and this is the initial location of all life now on earth. The landing of the ark on the mountaintop was not the end of the flood. All remained inside the ark until it was determined that land was a safe place on which to live.

Book of Genesis Chapter 8

8:5 And the waters decreased continually until the tenth month: in the tenth month, on the first day of the month, were the tops of the mountains seen.
8:6 And it came to pass at the end of forty days, that Noah opened the window of the ark which he had made:
8:7 And he sent forth a raven, which went forth to and fro, until the waters were dried up from off the earth.
8:8 Also he sent forth a dove from him, to see if the waters were abated from off the face of the ground,
8:9 But the dove found no rest for the sole of her foot, and she returned unto him into the ark, for the waters were on the face of the whole earth: then he put forth his hand, and took her, and pulled her in unto him into the ark.
8:10 And he stayed yet other seven days, and again he sent forth the dove out of the ark,

Book of Genesis Chapter 8

8:11 And the dove came in to him in the evening, and, lo, in her mouth was an olive leaf pluckt off: so Noah knew that the waters were abated from off the earth.
8:12 And he stayed yet other seven days, and sent forth the dove, which returned not again unto him any more.
8:13 And it came to pass in the six hundredth and first year, in the first month, the first day of the month, the waters were dried up from off the earth: and Noah removed the covering of the ark, and looked, and, behold, the face of the ground was dry.
8:14 And in the second month, on the seven and twentieth day of the month, was the earth dried.

Noah sent out a raven and a dove in search of dry land. The dove returns first time without finding dry land and we are not told about the raven. The second releasing after seven more days, the dove it returns with the leave of an olive tree. Being covered with water for at least 197 days has destroyed all things on earth. The flood was 150 days and they waited on the mountain for 40 days and it was 7 days before sending the dove out again, and there are other days and times scattered in to be disruptive to the calculation but at least it was 197 days. Now everything is gone and the earth is starting over yet the dove returns with an olive leaf, which would suggest an olive tree. Has God been creating again beyond what the story tells? The earth is dry and the people and animals from the ark may return to normal life. Vegetation must have returned quickly for the herbivores to maintain life. The carnivores must have gone hungry waiting on the following generations of prey to avoid eliminating the species on which they prey. Nonetheless, it must have been great to get back to earth.

Book of Genesis Chapter 8

8:15 And God spake unto Noah, saying,
8:16 Go forth of the ark, thou, and thy wife, and thy sons, and thy sons'
wives with thee.
8:17 Bring forth with thee every living thing that is with thee, of all
flesh, both of fowl, and of cattle, and of every creeping thing that
creepeth upon the earth, that they may breed abundantly in the earth,
and be fruitful, and multiply upon the earth.
8:18 And Noah went forth, and his sons, and his wife, and his sons'
wives with him:
8:19 Every beast, every creeping thing, and every fowl, and whatsoever
creepeth upon the earth, after their kinds, went forth out of the ark.
8:20 And Noah builded an altar unto the LORD, and took of every clean
beast, and of every clean fowl, and offered burnt offerings on the altar.
8:21 And the LORD smelled a sweet savour, and the LORD said in his
heart, I will not again curse the ground any more for man's sake, for the
imagination of man's heart is evil from his youth, neither will I again
smite any more every thing living, as I have done.
8:22 While the earth remaineth, seedtime and harvest, and cold and
heat, and summer and winter, and day and night shall not cease.

Verse 20 "And Noah builded an altar unto the LORD, and took of every clean beast, and of every clean fowl, and offered burnt offerings on the altar." presents a paradox. The LORD has Noah save the clean animals from the flood and then allows Noah to sacrifice at least one of every clean animal and bird. This was the reason for directing Noah to take multiples of every clean animal and fowl. But is this not absurd? God was going to kill them with the flood. Instead, He has Noah save them and then sacrifice them on the altar. The LORD and God are used interchangeably and assuming they are the same deity for a monotheistic religion. So, God's lust of killing obscures the ludicrousness of His actions. God is so pleased with the smell of burning flesh He changes his heart to never again destroy every living thing. God does not set limits on the amount of destruction He might employ. It could be interpreted as He could use nuclear weapons and destroy every living thing except cockroaches and not reach the limit of destroying every living thing. The statements of not cursing the earth and the days and seasons will not cease are too vague to be scientific. They create part of the source of disagreement between religion and science.

Verse 22 is interesting. As long as the earth exists, cycles will not cease. In Joshua 10:12, Joshua commands the sun and moon to stand still for a day. This would interrupt the day and night God promised would not cease. How can this be? Actually, the story of Joshua is adapted from the Theban religion of Egypt where the sun and moon are the deities and by commanding them to stand still, Joshua is asking them not to intercede in his battle. When adapted to the Bible they were changed from deities to just the sun and moon. So, there is an interpretation problem of changing deities in the stories but the sun and moon did not stand still and the earth did not stop rotating for a day. As for the agricultural cycles, they will remain but it is very likely they will relocate with global warming. The agricultural areas will die and have already shown great movements due to the changes in climate but they have not ceased globally. This may be of little comfort to those who starve due to their lands ceasing to support crops, but the seedtime and harvest have not ceased globally. So, it appears the word of God is good globally but not necessarily locally.

Just a extra note on the flood is added although not part of the Bible covered here. Two Columbia University marine geologists in 1996 released finding that the Bible's Great Flood was plagiarized tales from the Black Sea flood. The Black Sea catastrophic flood took place in about 7,600 years ago, long before the creation of the Biblical story. This is independent of analyzing the Bible's story but gives some evidence that it may have been from tribal knowledge passed down.

Questions over Chapter 12

1. The ark landed on mountains in Ararat
   a. after 40 days.
   b. because it was the highest mountains of the time.
   c. on the 17th of the 7th month.
   d. and we are not told why.

2. Ararat is
    a. mountain we are not told its location and it has never been found.
    b. in Israel.
    c. East of the Garden of Eden.
    d. in Egypt.
3. After landing on Ararat
    a. the people and animals return to dry land.
    b. Noah waited 40 days and sends out a Raven and a Dove.
    c. Noah sacrificed one of the clean animals.
    d. life returned to normal.
4. The dove returned with an Olive leaf.
    a. after being released a second time a week after the first failure.
    b. which is amazing that a tree grew so quickly.
    c. which revealed that there was dry land.
    d. all the above.
5. Noah built an alter on dry ground
    a. to praise God for saving them.
    b. to start the new life.
    c. to sacrifice one of every clean animal.
    d. because God directed him to do so.
6. God is pleased with the sacrifice and
    a. promises to never again curse the earth for man's sake and never kill everything living.
    b. promises to never flood the earth again.
    c. promises to never destroy the earth.
    d. has to be the reason for taking extra clean animals over the not clean.
7. God promises not to curse the ground again
    a. because God does not want to start over again.
    b. for man's sake.
    c. by using floods.
    d. all the above.
8. God says he will not kill every living thing
    a. because Noah's heroic efforts.
    b. because the LORD smelled a sweet savour (savoir) of the burnt offerings on the altar.
    c. because the flood was worse than expected.

d. we are not told why.

Probability of statements in Chapter 12

Instructions: Assign a number to the plausibility of each statement.
(1) For a total contradiction to a statement.
(2) For an implausible statement.
(3) For a possible and maybe even practical statement.
(4) For a plausible statement.
(5) For a confirmed or verifiable statement.

1. What is the probability that Ararat exist and has never been found? (1 – 5)?
2. What is the probability that Noah was able to repopulate the earth's animal population from a ship slightly bigger than a 3 level football field? (1 – 5)?
3. What is the probability that Noah could supply food and water to all animals for more than 197 days? (1 – 5)?
4. What is the probability that an olive tree could survive under water more than 197 days. (1 – 5)?
5. Noah had to take extra animals to have some to sacrifice after returning to dry land. What is the probability that this is consistent with a benevolent God? (1 – 5)?
6. What is the probability that God will not kill every living thing has a clear meaning? (1 – 5)?
7. Floods were feared as they brought death but also it was known that floods brought nutrients that made the earth fertile and provided life. This knowledge was rooted in Egypt and Mesopotamia. What is the probability that the Bible writers and scribes did not know this and wrote the Flood Story independent of this knowledge from other lands? (1 – 5)?

Chapter 13: Repopulating the World

Book of Genesis Chapter 9

9:1 And God blessed Noah and his sons, and said unto them, Be fruitful, and multiply, and replenish the earth.
9:2 And the fear of you and the dread of you shall be upon every beast of the earth, and upon every fowl of the air, upon all that moveth upon the earth, and upon all the fishes of the sea, into your hand are they delivered.
9:3 Every moving thing that liveth shall be meat for you, even as the green herb have I given you all things.
9:4 But flesh with the life thereof, which is the blood thereof, shall ye not eat.
9:5 And surely your blood of your lives will I require, at the hand of every beast will I require it, and at the hand of man, at the hand of every man's brother will I require the life of man.
9:6 Whoso sheddeth man's blood, by man shall his blood be shed: for in the image of God made he man.

God sets out new rules for the new beginning of earth. First is for Noah and his sons to be fruitful (prolific) and repopulate the world. God tells Noah and family that all creatures will fear them and they are delivered into their hands. This verse instructs that man may use animals but it also denotes man as a protector of animals. With the number of species that have become extinct since this verse was written it appears we have not done well in our protective role and predictions are that our performance on this will get worse. But also God has changed from making humans strict vegetarians to allowing certain meats to be consumed. The writers are being a bit cryptic in their explanations and these verses are not clear. The inerrant word of God would be expected to be clear and precise. The meanings of verses 9:3 through 9:6 are varied from the interpretation that blood contains the source of life and blood must be drained from all flesh before consuming to this could be construed as to disallow eating anything containing life. These verses may be as simple as to disallow eating any meat that used blood to sustain live. This may have been if people believed that some fish and very small animals did not require blood for life, as they could not detect it. Verse 9:5 is also a bit bazaar. It implies God wishes sacrifices to continue as He enjoys bloodletting. Contrary to this implication, verse 9:5 may be construed, as a warning that bloodletting or killing will be punished by bloodletting or killing, whether it is from animals or man. But the last part of Genesis 9:5 implies God requires the life of a man. Common belief is not to associate this with human sacrifice and tries to draw a different meaning. It may be taken as part of the warning not to take blood. The interpretation can be that all men are required to provide a blood sacrifice and that may be taken from beast.

The next verse 9:6 seems to reiterate this denial of human sacrifice as God sets the punishment for spilling the blood of a man with the loss of blood from the perpetrator. This goes along with the doctrine of an eye for and eye which comes later in the Bible and is generally the perception by most readers. It sets the tone of punishment for Bible readers and tends to tie these verses together as a warning rather than the reiteration of requiring animal sacrifices. Again, in this verse, God reminds us He has created man in his image. The thought pattern seems to be a bit random here. The writers have a lot to say and try to force into a few verses rather than presenting it logically in a more organized form.

Book of Genesis Chapter 9

9:7 And you, be ye fruitful, and multiply, bring forth abundantly in the earth, and multiply therein.
9:8 And God spake unto Noah, and to his sons with him, saying,
9:9 And I, behold, I establish my covenant with you, and with your seed after you,
9:10 And with every living creature that is with you, of the fowl, of the cattle, and of every beast of the earth with you, from all that go out of the ark, to every beast of the earth.
9:11 And I will establish my covenant with you, neither shall all flesh be cut off any more by the waters of a flood, neither shall there any more be a flood to destroy the earth.
9:12 And God said, This is the token of the covenant which I make between me and you and every living creature that is with you, for perpetual generations:
9:13 I do set my bow in the cloud, and it shall be for a token of a covenant between me and the earth.
9:14 And it shall come to pass, when I bring a cloud over the earth, that the bow shall be seen in the cloud:
9:15 And I will remember my covenant, which is between me and you and every living creature of all flesh, and the waters shall no more become a flood to destroy all flesh.
9:16 And the bow shall be in the cloud, and I will look upon it, that I may remember the everlasting covenant between God and every living creature of all flesh that is upon the earth.
9:17 And God said unto Noah, This is the token of the covenant, which I have established between me and all flesh that is upon the earth.
9:18 And the sons of Noah, that went forth of the ark, were Shem, and Ham, and Japheth: and Ham is the father of Canaan.

Starting in 9:7 God changes tone to benevolent and caring. He expresses to Noah and sons that the earth is new and provided for them to go and multiply to refill the earth. God continues with His promise to never destroy all living things by flood. God said His promise is designated by the bow (what we refer to as a rainbow) in the sky. God said this covenant or promise would perpetuate through following generations. Note God did not say He would not selectively use floods, but only that He would not use floods to kill everything and every person. Including a purpose for the rainbow may be a hint that the writers of the Torah or Pentateuch did not understand the physical phenomena that created rainbows. The physics of light and the cause of rainbows are now understood. Rainbows are created when light intersect a high density of random spheres causing diffusion into the color spectrum. Without the understanding of light or interference of waves and diffraction, the rainbow would be mystical and tempting to project as a symbol from God. Rainbows are seen when looking at a storm while looking away from the sun. The directions to produce the observation of a rainbow are most likely after a storm has passed. Since the rainbow appears following the storm or when a storm has passed a location, it is visible during calm weather and would then suggest the peace after a storm. Being a natural symbol for the calm after a storm and not understanding the physics causing the light diffraction would lead people to the conclusion of a rainbow being a mystical representation produced by God for promising peace or security from the storm. This would seem to be a very likely scenario for writing the promise from God in Genesis 9:11 through 9:17.

So, Noah and sons and their families left the ark. A common practice at the time of creating the Bible was for land areas to be named after a ruler or leader, sometimes referred to as the King. Verse 9:18 is a little different in that it points out that Ham was the father of Canaan, and Canaan is the name of the land Jews will return to after exile in Egypt.

Book of Genesis Chapter 9

9:19 These are the three sons of Noah: and of them was the whole earth overspread.
9:20 And Noah began to be an husbandman, and he planted a vineyard:
9:21 And he drank of the wine, and was drunken, and he was uncovered within his tent.
9:22 And Ham, the father of Canaan, saw the nakedness of his father, and told his two brethren without.
9:23 And Shem and Japheth took a garment, and laid it upon both their shoulders, and went backward, and covered the nakedness of their father, and their faces were backward, and they saw not their father's nakedness.
9:24 And Noah awoke from his wine, and knew what his younger son had done unto him.
9:25 And he said, Cursed be Canaan, a servant of servants shall he be unto his brethren.
9:26 And he said, Blessed be the LORD God of Shem, and Canaan shall be his servant.
9:27 God shall enlarge Japheth, and he shall dwell in the tents of Shem, and Canaan shall be his servant.
9:28 And Noah lived after the flood three hundred and fifty years.
9:29 And all the days of Noah were nine hundred and fifty years: and he died.

The writers seem to wander from the creation to moralizing. The story departs from repopulating the world to Noah's planting a vineyard and drinking excessively. This appears to be included as an admonition from the writers to the abuse of wine. Noah was 600 years old at the start of the flood. He has come through the flood and planted a vineyard and produced wine. Even if the vineyard grew quickly and he produced wine from the first crops this had to be a few years. So, Noah is at least 603 years old. When a man has lived 603 years he should be allowed to partake in alcohol and even to excess without being reprimanded. If Noah lived another 347 years as reported then apparently the drinking to excess did not damage his liver as would be expected.

Noah coming out from under the influence of the wine somehow knew Ham had viewed him naked and his brothers had not. This was very perceptive to say the least. To curse Ham and his son Canaan and the land seems a bit sever for viewing Noah naked when Noah's drunken state placed him in the vulnerable condition for observation. This point of the story does not appear to be about moral values based on logic and reasons but to create a cause for a curse on Canaan, the land of Ham. Since Ham had discovered Noah nude and the other sons made an effort not to view their father naked, Ham and his land were cursed devoid of logic and reason.

Questions over Chapter 13

1. God has ordered Noah and his sons to be fruitful
    a. to never sin again.
    b. go forth and start new cities.
    c. and multiply and replenish the earth.
    d. all the above
2. God tells Noah and family that all animals fear man
    a. and man can do as he pleases.
    b. and are delivered into his hands, which may also imply protecting them.
    c. so watch for animal attacks.
    d. so they should be sacrificed regularly.
3. God says all moving living things are meat
    a. with a cryptic exception for when an animal is alive and its life is supported by blood.
    b. except clean animals.
    c. except animals that bleed.
    d. as long as they are eaten with herbs of the field.
4. Verses 9:3 through 9:6
    a. allows animal and human sacrifice.
    b. are clear to not kill.
    c. allow the consumption of meat.
    d. are cryptic but likely imply a warning for blood taking.
5. God reminds us that
    a. we are in the image of God.
    b. taking a mans blood will result in the same, as punishment.
    c. is still not clear on the previous verses but does imply tying them together as a warning.
    d. all the above.

6. Verse 9:7
    a. continues the warning.
    b. seems to change toward benevolence.
    c. is still unclear.
    d. all the above.
7. God makes a covenant with Noah
    a. and all the population to follow.
    b. with all animals.
    c. to never eliminate all life again.
    d. all the above.
8. The bow in the clouds
    a. is due to rain.
    b. in the creations story is a token of God's covenant.
    c. was certainly known to follow storms.
    d. started the story of locating riches at the end of the rainbow.
9. Noah plants a vineyard and produces wine
    a. which leads to his drunken state.
    b. to provide for his son's families.
    c. as the first winery.
    d. which leads to his death.
10. Ham viewed his father naked
    a. because Ham was drunk.
    b. and warned his brothers.
    c. and after he uncovered Noah.
    d. but Noah was passed out and did not know.
11. Noah awoke from his drunken state
    a. and knew that Ham had seen him naked even though he was now covered.
    b. cursed Ham and his descendants for his sin.
    c. and knew Shem and Japheth were innocent of the sins committed by Ham of seeing his father naked.
    d. all the above.
12. After the drunken stupor led to cursing Ham, Ham's son Canaan, and their descendants and their lands Noah
    a. died.
    b. sobered up.

c. lived to be 950 years old.
d. helped Ham and Canaan establish Israel.

Probability of statements in Chapter 13

Instructions: Assign a number to the plausibility of each statement.
(1) For a total contradiction to a statement.
(2) For an implausible statement.
(3) For a possible and maybe even practical statement.
(4) For a plausible statement.
(5) For a confirmed or verifiable statement.

1. What is the probability that seeing a man naked was worthy of a curse to all future generations? (1 – 5)?
2. What is the probability that writers of the Bible understood rainbows and the phenomena causing them? (1 – 5)?
3. A possible interpretation of the God saying all animals fear man and they are delivered into his hands, could be that man is also the protector of all animals. In recent history, Christians have opposed actions of environmentalists and animal protectionists. What is the probability that Christians understand this verse? (1 – 5)?
4. The creation story instructs man to eat of the herbs of the field. Verses 9:3 and 9:4 could easily be interpreted as man should eat meat or flesh of animals. What is the probability that instructions on the use of animals for food are clear? (1 – 5)?
5. The flow of the creation story appears to go from storytelling to warnings to instructions and then to expressing benevolence and encouragement to populate the world and be the protector. With the random changes in prospective and the story line, what is the probability that the consistencies were written by a super intelligent being or at least guided the writing of the Bible? (1 – 5)?
6. What is the probability that Noah could have awoken hung-over and known that Ham had

viewed him naked and that Shem and Japheth had not? (1 – 5)?

7. Common sense would evaluate the creators of the creation stories as inserting their opinions on the morals of drinking to excess and the immorality of nudity and the punishment of bloodshed for causing bloodshed. What is the probability that these stories are actually divinely created as the inerrant word of God? (1 – 5)?

Chapter 14: World's Population Diversity

Chapters 10 and 11 of the Book of Genesis require a lot of reading for little information. They are included for completeness but only recommended as a historical interest not scientific. This is a science textbook and the relevant information is listed following the copy of chapters 10 and 11.

Book of Genesis Chapter 10

10:1 Now these are the generations of the sons of Noah, Shem, Ham,
and Japheth: and unto them were sons born after the flood.
10:2 The sons of Japheth, Gomer, and Magog, and Madai, and Javan,
and Tubal, and Meshech, and Tiras.
10:3 And the sons of Gomer, Ashkenaz, and Riphath, and Togarmah.
10:4 And the sons of Javan, Elishah, and Tarshish, Kittim, and
Dodanim.
10:5 By these were the isles of the Gentiles divided in their lands, every
one after his tongue, after their families, in their nations.
10:6 And the sons of Ham, Cush, and Mizraim, and Phut, and Canaan.
10:7 And the sons of Cush, Seba, and Havilah, and Sabtah, and
Raamah, and Sabtechah: and the sons of Raamah, Sheba, and Dedan.
10:8 And Cush begat Nimrod: he began to be a mighty one in the earth.
10:9 He was a mighty hunter before the LORD: wherefore it is said,
Even as Nimrod the mighty hunter before the LORD.
10:10 And the beginning of his kingdom was Babel, and Erech, and
Accad, and Calneh, in the land of Shinar.
10:11 Out of that land went forth Asshur, and builded Nineveh, and the
city Rehoboth, and Calah,
10:12 And Resen between Nineveh and Calah: the same is a great city.
10:13 And Mizraim begat Ludim, and Anamim, and Lehabim, and
Naphtuhim,
10:14 And Pathrusim, and Casluhim, (out of whom came Philistim,)
and Caphtorim.
10:15 And Canaan begat Sidon his first born, and Heth,
10:16 And the Jebusite, and the Amorite, and the Girgasite,
10:17 And the Hivite, and the Arkite, and the Sinite,
10:18 And the Arvadite, and the Zemarite, and the Hamathite: and
afterward were the families of the Canaanites spread abroad.

Book of Genesis Chapter 10

10:19 And the border of the Canaanites was from Sidon, as thou comest to Gerar, unto Gaza, as thou goest, unto Sodom, and Gomorrah, and Admah, and Zeboim, even unto Lasha.
10:20 These are the sons of Ham, after their families, after their tongues, in their countries, and in their nations.
10:21 Unto Shem also, the father of all the children of Eber, the brother of Japheth the elder, even to him were children born.
10:22 The children of Shem, Elam, and Asshur, and Arphaxad, and Lud, and Aram.
10:23 And the children of Aram, Uz, and Hul, and Gether, and Mash.
10:24 And Arphaxad begat Salah, and Salah begat Eber.
10:25 And unto Eber were born two sons: the name of one was Peleg, for in his days was the earth divided; and his brother's name was Joktan.
10:26 And Joktan begat Almodad, and Sheleph, and Hazarmaveth, and Jerah,
10:27 And Hadoram, and Uzal, and Diklah,
10:28 And Obal, and Abimael, and Sheba,
10:29 And Ophir, and Havilah, and Jobab: all these were the sons of Joktan.
10:30 And their dwelling was from Mesha, as thou goest unto Sephar a mount of the east.
10:31 These are the sons of Shem, after their families, after their tongues, in their lands, after their nations.
10:32 These are the families of the sons of Noah, after their generations, in their nations: and by these were the nations divided in the earth after the flood.

Book of Genesis Chapter 11

11:1 And the whole earth was of one language, and of one speech.
11:2 And it came to pass, as they journeyed from the east, that they found a plain in the land of Shinar, and they dwelt there.
11:3 And they said one to another, Go to, let us make brick, and burn them thoroughly. And they had brick for stone, and slime had they for morter.
11:4 And they said, Go to, let us build us a city and a tower, whose top may reach unto heaven, and let us make us a name, lest we be scattered abroad upon the face of the whole earth.
11:5 And the LORD came down to see the city and the tower, which the children of men builded.
11:6 And the LORD said, Behold, the people is one, and they have all one language, and this they begin to do: and now nothing will be restrained from them, which they have imagined to do.
11:7 Go to, let us go down, and there confound their language, that they may not understand one another's speech.

Book of Genesis Chapter 11

11:8 So the LORD scattered them abroad from thence upon the face of all the earth: and they left off to build the city.
11:9 Therefore is the name of it called Babel, because the LORD did there confound the language of all the earth: and from thence did the LORD scatter them abroad upon the face of all the earth.
11:10 These are the generations of Shem: Shem was an hundred years old, and begat Arphaxad two years after the flood:
11:11 And Shem lived after he begat Arphaxad five hundred years, and begat sons and daughters.
11:12 And Arphaxad lived five and thirty years, and begat Salah: **11:13**
And Arphaxad lived after he begat Salah four hundred and three years, and begat sons and daughters.
11:14 And Salah lived thirty years, and begat Eber:
11:15 And Salah lived after he begat Eber four hundred and three years, and begat sons and daughters.
11:16 And Eber lived four and thirty years, and begat Peleg:
11:17 And Eber lived after he begat Peleg four hundred and thirty years, and begat sons and daughters.
11:18 And Peleg lived thirty years, and begat Reu:
11:19 And Peleg lived after he begat Reu two hundred and nine years, and begat sons and daughters.
11:20 And Reu lived two and thirty years, and begat Serug:
11:21 And Reu lived after he begat Serug two hundred and seven years, and begat sons and daughters.
11:22 And Serug lived thirty years, and begat Nahor:
11:23 And Serug lived after he begat Nahor two hundred years, and begat sons and daughters.
11:24 And Nahor lived nine and twenty years, and begat Terah:
11:25 And Nahor lived after he begat Terah an hundred and nineteen years, and begat sons and daughters.
11:26 And Terah lived seventy years, and begat Abram, Nahor, and Haran.
11:27 Now these are the generations of Terah: Terah begat Abram, Nahor, and Haran, and Haran begat Lot.
11:28 And Haran died before his father Terah in the land of his nativity, in Ur of the Chaldees.
11:29 And Abram and Nahor took them wives: the name of Abram's wife was Sarai, and the name of Nahor's wife, Milcah, the daughter of Haran, the father of Milcah, and the father of Iscah.
11:30 But Sarai was barren, she had no child.
11:31 And Terah took Abram his son, and Lot the son of Haran his son's son, and Sarai his daughter in law, his son Abram's wife, and they went forth with them from Ur of the Chaldees, to go into the land of Canaan, and they came unto Haran, and dwelt there.

Book of Genesis Chapter 11

11:32 And the days of Terah were two hundred and five years: and Terah died in Haran.

These chapters of 10 and 11 appear to be the writers' endeavor to add credibility by listing the succession of generations from Noah to Abraham. As the earth is repopulated from the decedents of Noah, the writers perceive a major problem at this point in the story. The world even known by the writers was more diversified than would result in descendants from one man and woman. There was an attempt to rationalize the difference in languages in versus 10:20 and 10:31 where it is said Noah's descendants spread through the Middle Eastern area and spoke in their own tongues in their own countries and nations.

Now the first verse of Chapter 11, says the earth was of one tongue. This is a contradiction the writers have to resolve. The story of Babel was written for this purpose. Several things are peculiar about the story of Babel. First is that the city of Babel has never been located or identified or any ruins that may be associated with the previous existence. Second is that God could be concerned about a tower far less significant than the pyramids of Egypt. The reference as a tower implies proportions of height to base are a larger ratio than the pyramids of Giza. A tower would indicate near vertical walls. This brings in the structural limitations of building with bricks. Given the benefit of doubt that the ancient people could locate superior clay for optimum strength and they empirically developed perfected firing, the bricks they produced would still limit the structure to less than 600 feet[13] in height. A 600-foot structure would have stressed the limits of feasibility, but still would not penetrate heaven. It would be less height than many buildings of today. God either accepted the new practice and allows modern buildings to penetrate heaven or the tower of Babel is a fable to attempt to explain the diversity of the world's population. The story of Babel attempts to explain the difference in language but does not address the biological differences of Africa and the Orient and the Americas. Writers of the Pentateuch may not have known of natives from Tanzania or Peru. The omission of differences in people cannot be based on the writers of the Bible being non bigoted, since they very are careful to distinguish between every sect and group in their known area of Canaan, North Africa and Mesopotamia. The difference in people is denoted through the Bible to ensure us the writers were quite cognizant of the slightest differences in ethnicity. The tower of Babel does not explain the physical differences in people.

The completion of the creation and populating the world ends with continuing the listing of generations leading to Lot and Abram, later referred to Abraham. The development of the Old Testament stories continues from this heritage.

Charts are to assist in understanding the descendants from Noah. Only the sons or males are listed with one exception. Each chart is for a son of Noah, Japheth, Ham and Shem. These family trees were included to show the spread across the known world and the descendants of Noah following the Tower of Babel disrupting language and cultures in all known lands.

Heirs of Japheth. Japheth son of Noah

- **Gomer**
  - **Ashkenaz**
  - **Riphath**
  - **Togarmah**
- **Magog**
- **Madai**
- **Javan**
  - **Elishah**
  - **Tarshish**
  - **Kittim**
  - **Dodanim**
- **Tubal**
- **Meshech**
- **Tiras**

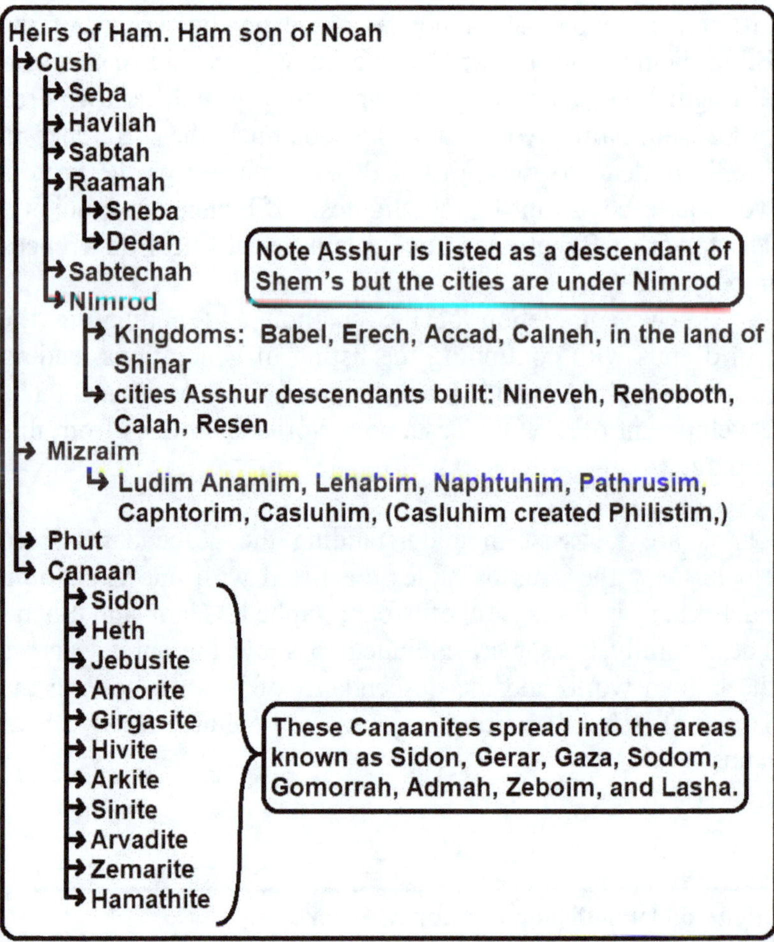
Heirs of Ham. Ham son of Noah
Cush
Seba
Havilah
Sabtah
Raamah
Sheba
Dedan
Sabtechah
Nimrod
Kingdoms: Babel, Erech, Accad, Calneh, in the land of Shinar
cities Asshur descendants built: Nineveh, Rehoboth, Calah, Resen
Note Asshur is listed as a descendant of Shem's but the cities are under Nimrod
Mizraim
Ludim Anamim, Lehabim, Naphtuhim, Pathrusim, Caphtorim, Casluhim, (Casluhim created Philistim,)
Phut
Canaan
Sidon
Heth
Jebusite
Amorite
Girgasite
Hivite
Arkite
Sinite
Arvadite
Zemarite
Hamathite
These Canaanites spread into the areas known as Sidon, Gerar, Gaza, Sodom, Gomorrah, Admah, Zeboim, and Lasha.

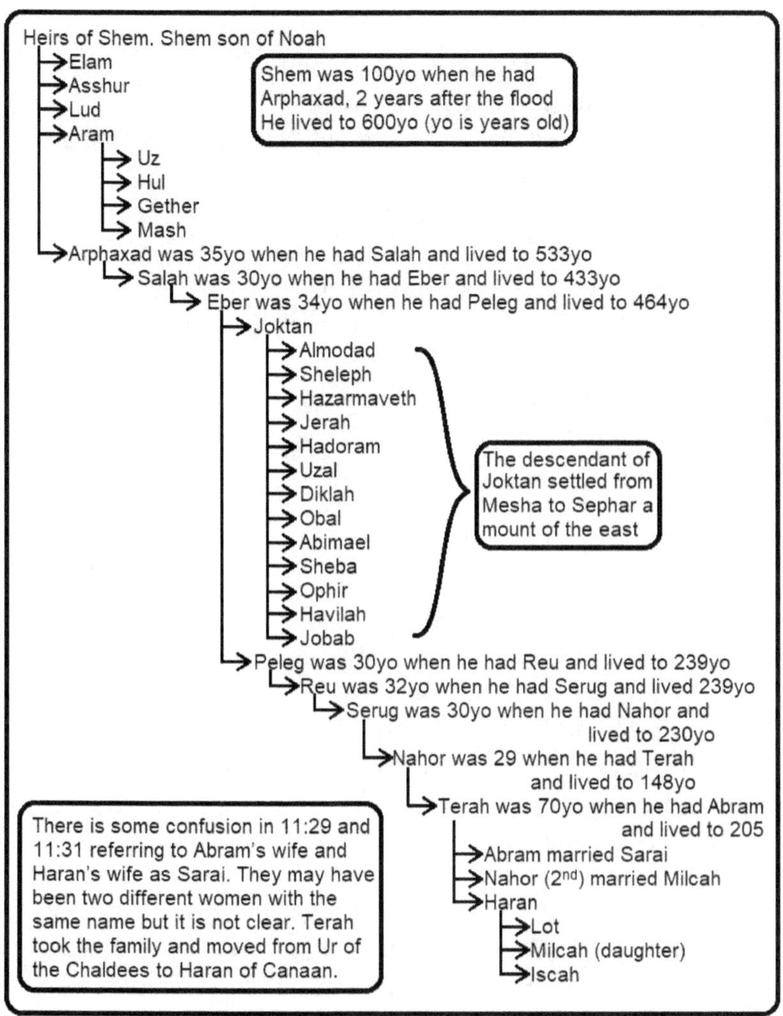

Questions over Chapter 14

1) Chapter 10 says the descendants of Noah spread through the Middle East
   a) and established their own countries and nations.
   b) spoke in their own languages.
   c) established kingdoms often named after descendants.
   d) all the above.

2) Chapter 11:01 says
   a) each Kingdom had its own language.
   b) the world spoke with one tongue.
   c) the language was the same but dialects were change.
   d) none of the above.
3) The ancestry heritage of people
   a) is listed from Noah on.
   b) cannot be proven or disproved.
   c) may be included by the writers to add creditability.
   d) all the above.
4) The tower of Babel
   a) had to exceed the height of the pyramids.
   b) had to be less than 600 feet.
   c) had to be built on top of a mountain to be useful.
   d) was locate west of the Dead Sea.
5) The story of the tower of Babel
   a) explains the different races around the world.
   b) attempts to explain the different languages in area where the Bible was written.
   c) is why in general women are smaller than men.
   d) is the writers way to show the Jews could build better than the Egyptian pyramids.
   e) to make the story complete.
   f) we are not told
6) Genesis Chapter3 verse 6 limits man's life to 120 years
   a) but the flood changed this to again exceed 120 years.
   b) but the life spans of chapters 10 & 11 exceed this.
   c) which was only a threat.
   d) which has held ever since.

Probability of statements in Chapter 14

Instructions: Assign a number to the plausibility of each statement.
(1) For a total contradiction to a statement.
(2) For an implausible statement.
(3) For a possible and maybe even practical statement.
(4) For a plausible statement.
(5) For a confirmed or verifiable statement.

1. What is the probability that the nations of the world spoke with one language when the descendants spread out developed their own languages (1 – 5)?
2. The writers of the Bible construct ten generations leading to Abram with no record keeping. With today's records in public and government offices it is difficult to trace more than a few generations back. What is the probability that the heritage leading to Abram is accurate (1 – 5)?
3. What is the probability that the tower of Babel ever existed, was built after the pyramids and should have been more spectacular, and left no trace (1 – 5)?
4. What is the probability that God was irritated enough from the building of the tower to disrupt language, culture, and society by preventing people from organizing (1 – 5)?
5. What is the probability that the human race spread and instantly diversify worldwide from God's anger (1 – 5)?
6. What is the probability that the writers of chapters 10 & 11 were aware of chapter 6 of Genesis that limited life to 120 years (1 – 5)?
7. What is the probability that writing the linage from Noah to Abraham provides credibility to Genesis (1 – 5)?
8. In many verses through Genesis the wording is obscure and difficult to derive the meaning and in many case the ambiguity can result in several interpretations. Making a decision from the first eleven books of Genesis, what is the probability that these are the inerrant word of God (1 – 5)?

Chapter 15: Intelligent Design is not a Science

There are two reasons for analyzing Intelligent Design in science classes, which requires an explanation before starting. ID[14] is a recent creation of the religious conservatives and evangelists as an attempt to displace natural science studies in public schools. Their motivation for this is that science contradicts their religion and the Bible. ID was conceived after many years of trying to force the teaching creationism have yielded minimal results. The movements to teach creationism has increased in importance in the vision of the Christians since Charles Darwin published *On the Origin of Species* in 1859. IDers[15] are trying a new approach using semantics to rename creationism and separate ID from religion in hopes of getting it added to public school curriculums. IDers labeled their beliefs of their view of creationism as science and propose ID be taught in lieu of biological science and eliminate all teaching of evolution. Textbooks on ID have been written for this purpose. So far this has failed. As an alternative IDers are proposing as a minimum to force high school science teachers to offer ID as an alternative in science classes to factual science. IDers have proposed that ID be taught parallel with biological science to provide students a choice. Until now this approach has failed to be effective. IDers in several states have achieved getting laws passed that require a warning label be put in all biological science books that states, "This textbook contains material on evolution. Evolution is a theory, not a fact, regarding the origin of living things. This material should be approached with an open mind, studied carefully and critically considered." IDers have forced through laws in some states that weaken the teaching of biological science. IDers have succeeded in getting laws that require teachers to accept any answer to any question in science if the student feels their religion is threatened. These laws are nebulous and in a more practical sense allow the student to not study and use religion as an excuse. The stated purpose was to allow students the right to hold the personal opinion of a creation. The effect is that students can avoid being graded for absurd answers. State legislators are yielding to IDers and the ID movement is gaining momentum. At the time of publishing this book, only a small percentage of students nationwide were being forced to accept ID by the law as a science course. But the intent of the religious right is to force all students nationwide to be taught ID as the science of life's origins. Even though ID is not science it is creeping into science classes, and with this plague infecting our science programs, this textbook approaches the subject by investigating ID truthfully and objectively. So, the first reason for teaching ID in science class is to truthfully and objectively analyze ID.

The second reason for teaching ID as a science class is to avoid introducing direct conflicts with religious doctrine in public schools. If ID is taught as creationism, then it is religion because ID has no evidence or quality as a science or references outside of religion. If ID is taught as a religion and then evolution is taught contradicting ID then there is a religious conflict, even under the guise of giving students a choice. ID could be legally studied under social studies or political science classes and analyzed as a social movement. This approach would not lessen the desire of IDers wanting ID to displace science. Also analyzing ID on a social level has the potential of creating the illusion attacking the character and intellect of those promoting it. The difference in analyzing ID as a science is that it is objective. Teaching ID as an unsupported science or analyzing ID as a social religious movement is subjective and may be construed as attack on religious beliefs. The analysis of ID as a science disproves the reasoning of the story. The analysis of ID as a social religious movement discredits the evangelists and proponents, which becomes a criticism of the dishonesty of the religion[16]. Analyzing ID as a social religious movement becomes an attack on religious beliefs. The Supreme Court ruling allows discussion of religions in a broad sense but this would be challenged if taken to the point of disproving religious tenants and focusing on Christian or Jewish religions. Social Studies analyzing of ID will be an attack on religion, which is not acceptable per Supreme Court rulings. Therefore, the analysis of ID must be a science review of the theory ID and avoid direct contradiction of the religious beliefs.

Creationism is straight from the Bible and cannot be separated from God and therefore has not been able to overcome the barrier of not teaching religion outside of church schools. Forcing creationism into public schools has not obtained the success proponents desired. Religious zealots created a new approach trying to conceal their God and the Bible but teaching the same philosophy to eliminate science courses that contradict their belief of creationism. IDers published a textbook *Of Pandas and People* attempting to introduce ID as a science class. The Bible does not mention ID so this approach was touted non-religious. ID is not science, it is an attempt to replace the creation in the Bible with a palatable replacement using a supernatural power to replace God. This is an attempt to gain entry into school curriculums replacing evolution with a 'secular' creation since the 'theory' does not mention their particular God. Acceptance has not been as overwhelming as planned because science teachers have recognized the deception, so IDers have proposed that ID be taught in parallel with evolution as a fair and balanced solution to teaching science. IDers failed to see one problem with this fair and balanced approach, it is wrong! Science is the study and organization of facts and truths, not a fair and balanced game to allow the student to select answers. Science is not an option of either faith and fables or difficult to comprehend facts to answer tough questions. IDers tried the approach of students should have the freedom to keep an open mind and be taught ID as well as evolution. IDers proposed that to keep an open mind student had to be taught both evolution and ID and that both were true and let student decide which suited them the best. The parallel approach was an entry attempt that also has had little success. IDers failed to recognize that science cannot teach the contradiction of both being true. It has not been a practice to teach students that 2+2=4 and 2+2=35. Science teachers and teachers' organizations have referred to this parallel approach as the Trojan Horse approach. IDers preferred the name the 'wedge' approach, which also reveals their strategy. Gaining entry with a 'generic superpower creating the world' contradicts teaching evolution and provides an opportunity to replace science with spiritual creation. Teaching that evolution is wrong and creation is the acceptable human history opens the door for proselytizing students and reduces academic conflicts with Christian religions.

ID is a recent creation by religious zealots starting to be pushed in the late 1980's. Several textbooks were published and attempts made to entice school boards to replace biological science books that refer to evolution or even the age of the earth or life forms. This would require replacement of all biological science books, which was opposed by science teachers and school boards. The goal of the new books was clear that it was not teaching science but protecting Christian tenants and not contradicting the Bible. ID contradicts science rather than supporting it or even coexisting with science. Science is facts and ID is religious based desires, which makes the two subjects not compatible. The fact that ID is not a science has not reduced the obsession with IDers of using it to replace science. ID has been debated in state legislatures and even national legislatures. Tom Delay[17] exploited the Columbine school shootings to promote ID by stating the tragedy was a direct result of teaching evolution in public schools. Tom Delay's statement was absurd but yet widely accepted. Tom Delay's statement replaced logic with emotion and thus swayed the less intelligent members of Congress. Legislators have shown their limitations of logic; therefore, Schools cannot rely on legal protection from IDers as religious tolerance is diminishing[18]. Revealing that ID cannot be compared to science and that it is not a science has prevented ID from being added to school curriculums until now. Not teaching ID has created emotional inflammatory rhetoric, controversy in political and religious arenas and there is no reason to believe it will diminish. So rather than continue conflicts about ID lets analyze it as a true science to eliminate the controversy.

Science requires proof. It is not possible to prove something does not or did exist. Therefore, it is not possible to prove that there was not an intelligent being that controlled the changes that created us and is controlling the changes that continue. No evidence of the theory that a superior being created the human race and all animals has been discovered. Science is based on fact and even if the idea of our being created by ID cannot be disproved, it can be shown there is no evidence of it. If there is no evidence of ID then it does not belong as a science and is relegated to faith and superstition. If it is shown that ID is faith or superstition it shows IDers have alternative agendas than teaching science. Also, students and teachers must be aware of the attacks on science. State laws mandating stickers be placed in science books stating, "This textbook contains material on evolution. Evolution is a theory, not a fact, regarding the origin of living things. This material should be approached with an open mind, studied carefully and critically considered." are harmful to science. This is harmful because it is wrong. Evolution is fact! Stating it is not fact is dishonest. It is like stating the theory of gravity is just a theory therefore gravity does not exist. Or the theory of space is just a theory and space does not exist[19]. Remember the Bible says God placed the firmament above the earth to hold the lights called stars. Space, gravity, and evolution are theories but that does not mean they are wrong. In science theory refers to subject that is still actively being investigated and understood. The second approach is getting laws passed to disallow questions or make any answer acceptable to science questions that students do not wish to answer. This is wrong. Acceptance of any answer would make grading simple, everybody passes with perfect scores. What could be better? Maybe if students that studied and gave the correct answers received good grades and if those that were not capable of providing the correct answer were failed. This pass/fail philosophy has functioned in the past. Also, student should realize that tests, besides grading students, are also learning experiences. A good test expresses what students know and also makes them think and learn.

ID is more complex than just the lack of understanding of science. Observation of our present world endorses a theory of ID. First of all because it works! Water freezes to make ice. Air, the mixture of gases we breathe remains a gas even at the coldest temperatures of our atmosphere. It is a wonderful world we in which we live. And it is difficult to imagine this world coming from chaos without intelligent input and control. We possess not only the mechanisms to support life, but mechanisms that create pleasures that enhance life. But the opposite is true as well. This world we have been given to live in provides disease, parasites, Tsunamis, hurricanes, earthquakes etc. If the world were created from ID, why would the designers include natural disasters, man-made disasters, parasites, and pestilent? If our world were designed by intelligence, it would be more pleasant to live in a world free of ticks, fleas, mosquitoes, flies and natural disasters. If the designers are given credit for creating the wonderful physical laws that create a pleasant environment, they have to be blamed for the physical laws that create an unpleasant and even deadly environment. A world full of inconveniences and harm is contradictory to a world created from ID. An ID world would not create human beings, which contain the millions of flaws we do. Humans have emotional, mental and physical flaws. How many people require sensory assistance? Eyeglasses or contacts and hearing aids are quite common. How many people require some form of surgery with in the first 120 years of life. Why do we require aids to control our reproduction instead of it being a conscience choice? Why are we susceptible to bacterial and viral infections? These are not consistent with a world created by a super intelligent power that would be capable of such tasks.

Does it matter whether you are taught that God created man or that man was created by ID or the truth that man evolved and the process of evolution has not stopped? Yes, if human life is to continue to develop and improve this is important. Understanding science and our history is more important now than it has ever been because humans have obtained the capability to destroy our environment and life support systems. Humans are the ultimate predator. Collective as a society we fear nothing and thus collective we respect nothing. Understanding our past and present can help us determine our future. A productive and positive future will only be generated by understanding science and history. We live in a technology driven society. This may not be a popular view but it is correct. Without technological advances the earth cannot support the human race, as it exists now and as the population grows it would be even more devastating to lose technology. Progress will not come from squelching science with barbaric myth and legends. Creating new myths to endorse old tenants is no less harmful. Human nature that has brought the human race to the brink of destruction has to be understood to develop the means to halt the destruction of life support systems of the earth. This textbook does not address the recovery from the development of human trends and policies that are destructive. This textbook points out that there is absolutely no scientific reasoning behind creationism or ID. When any person through any means or media tells you creationism or ID are science and true you must realize they are either dishonest or ignorant or possibly both. If God tells you creationism or ID are science and true, and you are not dreaming or hallucinating and He tells you direct face to face, you can believe it, if He can prove He is The one God. Be cautious of accepting the word of false Gods, there are many of them and even the first commandment warns us of this. This talk of God is science humor. It is inserted for enjoyment and lightens a serious subject. Do not take it seriously, it is not going to happen.

Analyzing and learning about ID and creation are not to placate the religious fundamentalists or political proponents. The correct and truthful knowledge of the human situation removes any doubt that biological sciences are important and the need is for the best knowledge that can be obtained. There is still research, discovery and analysis to be made in science fields and the burdens of conflict with religion need to be understood and met head-on and eliminated to improve or as minimum to sustain our standard of living.

Questions over Chapter 15

1. Intelligent design is
    a. a science that is too complex for humans to understand.
    b. the description of the creation shortened to avoid controversy.
    c. not a science but a religious desire.
    d. the complete explanation and that is all God wants us to know.
    e. none of the above.
2. ID
    a. is a recent invention.
    b. was developed to counter Darwin's *On the Origin of Species.*
    c. is a failing attempt to eliminate evolution and be independent of religion.
    d. all the above.
3. ID has to be analyzed in science because
    a. state legislators passed laws to force this.
    b. it is science.
    c. science examines the facts of the subject and not the people or purpose of those proponents of ID.
    d. state legislators passed laws to require stickers in science books.
4. A second reason to teach ID in science class
    a. to teach it honestly and objectively.
    b. because the ID movement is gaining legal support to force it into public schools.
    c. IDers supporting the teaching have alternative motives to teaching science.
    d. all the above.
5. Intelligent design is mutually acceptable to science and religion
    a. is incorrect because it is not acceptable and creates a division between science and religion.
    b. because it resolves all conflicts with creation and evolution.

c. because it explains all life without bringing religion into the public schools.
d. because science accept facts.

6. Intelligent design cannot be taught with evolution because
   a. there is not enough time.
   b. they are contradictory.
   c. students should not mix religion and life.
   d. science teachers do not like giving students a choice.
7. ID is a fair and balanced approach
   a. is a ploy used to disguise religion as science.
   b. and should be taught as such.
   c. because it gives both sides.
   d. as science should be.
8. Laws in some states prohibits failing answers to science test questions that
   a. do not include ID.
   b. use factual answers.
   c. students do not like.
   d. requires answer contrary to religious beliefs.
9. Intelligent design created a fantastic world
   a. where water and air freeze.
   b. is what proponents of ID wish to express and ignore less pleasant attributes.
   c. is why it should be taught.
   d. which proves the story.
10. Understanding biological science is important
    a. to get a good grade.
    b. because change has not stopped.
    c. because the world is more complex and man's role has changed to the ultimate predator.
    d. all the above and then some!

Probability of statements in Chapter 15

Instructions: Assign a number to the plausibility of each statement.

(1) For a total contradiction to a statement.
(2) For an implausible statement.
(3) For a possible and maybe even practical statement.
(4) For a plausible statement.
(5) For a confirmed or verifiable statement.

1. What is the probability that Intelligent design is fact though it is not mentioned the Bible? (1 – 5)?
2. What is the probability that Intelligent design provides a true answer that we were made perfect and have not changed? (1 – 5)?
3. What is the probability that Intelligent design provides a true answer of how humans were created? (1 – 5)?
4. What is the probability that Intelligent design is true because the evolution theory has missing links? (1 – 5)?
5. What is the probability that Intelligent design is true because TV evangelist say it is? (1 – 5)?
6. A political use of religion is to discredit environmentalists as alarmist trying to make money from scaring the public. This diverts attention from those destroying the environment. What is the probability that the earth will continue to support the population explosion without understanding of science? (1 – 5)?
7. What is the probability that understanding the beginning of humanity is frivolous, i.e., not important? (1 – 5)?

Chapter 16: Analyzing Intelligent Design

Analyzing ID starts with the ID's basic hypothesis. ID states human beings were created by a supernatural power, too complex and advanced for humans to understand. ID postulates that human beings did not evolve from lower animals. Continuing the basic premise is that ID created people and people did not evolve from lower animals. There is no justification for either statement but the fallacy of these hypotheses is what needed to be proven. Start with the tenants offered by ID.

The first claim for ID is that it is simple and straightforward. That ID is simple and straight forward cannot be denied, but that is not proof that ID's hypotheses are true. The point that human beings were created by a supernatural power is simple and the proponents say no evidence is required. Proponents of ID claim if human beings evolved from lower animals, then the mechanism of evolving needs to be understood which is too difficult. So, IDers claim this is proof of the ID theory that it is simple and understandable. This is not proof and it is not evidence. Just because it is simple does not mean it is proof. It does show that IDers are utilizing peoples' reluctance to investigate and think. IDers are using the fact that most people want the easy solution but that does not provide answers to scientific questions. Simplicity may make ID popular but not correct.

Analyzing the tediously slow process of life evolving over hundreds of millions of life cycles and hundreds of millions of generations of animals is very difficult and requires generations to develop the knowledge base. Much of the proof of mapping from animals to humans has been enhanced with DNA[20] testing and analysis, which is recent compared to religious tenets. The study of human development has taken place in a very short time compared to the millenniums of dogma passed down by religions. Creation as stated in the Bible or by ID has a tradition accumulated over several millenniums compared to a scientific explanation that has been started in the last 200 years. Obviously from the period the story of creation was created, it has to be simple.

Vulnerability of people to the simple answers and a history relating the simple answers to endorse religious concepts has focused IDers' effort to squelch science that opposes religious views.

One of the claimed proofs of ID is the observation that the present generation of humans is no different from our ancestors. We are like our great-great grandparents or as far back as we have photographs. Human beings differ little from the descriptions given in the earliest documentations, writings and drawings on parchment, cuneiform, and cave walls. Since man started recording history and events, changes in the human physiology have not been documented until recent times[21]. This history of presuming human genetics to be constant develops beliefs presumed to be fact by ID proponents. In fact, humans have been changing and continue to change but the changes are subtle and only discernible through thorough investigation.

Because change has been so gradual, careful scientific investigation is required to measure the changes. But change has occurred and is detectable when measured on the population collectively. The changes are small because they have occurred in the relatively short time period of the development of the human life. Humans have existed less than 0.002 percent of the earth's existence. So, change is small compared to the overall evolution of life. Scientist have determined that changes are actually occurring at a faster rate now than in the past 300,000 years. Forensic data on past fossils and generations show change in average weight, height, life span, information gathering and disseminating capability, preferences, tastes, travel, and many other parameters too extensive to list. The largest change has been humans' capability to destroy their own environment. Less than 200 years ago man could not destroy his environment, today we not only have the ability but also, we are executing it.

Change is measured on populations by the movement in the mean or average. Mental capacity and capability are difficult to measure. IQ tests are not reliable and have never been administered in a statically significant sample size. IQ test have notoriously favored the wealthy and better educated since the tests are written by the better educated. IQ test are not administered to large population and there is no record of past generations' intelligence levels. A comparison is required to determine if mental capabilities have changed but IQ tests are not the answer. The simple answer is "there is no change so there is no way to measure it" and dismiss the thought that intelligence changed in the last 300,000 years, but this is wrong. A second subtler problem arises with this in the age of humans on earth, which also creates a conflict of religious views that the earth is less than 10,000 years old. Religious leaders prefer the Biblical determinations of human existence, which varies between scholars, most suggest 4000 to 6000 years, but always less than 10,000 years. Fossil remains of Human beings have been dated as far back as 300,000 years. To avoid this conflict of the first humans, let's consider a very intelligent man living in a village of ten thousand years ago, the maximum age considered by some Biblical scholars. Picture explaining to him how to change the font in his emails or change the screen saver on his cell phone. Explain to the same man when it is preferable to sell stocks short over buying future options. This could be argued that the intellect to understand these concepts is environmental conditioning and not inherent. But intelligence is from both inherent and environmental conditions. Natural selection has fostered the growth of comprehension over the last nineteen thousand generations. Higher intelligence has led to higher survival rates and an increased probability of reproducing, at least until the introduction of birth control. Humans that had utilized technology and advanced with it have tended to have more surviving children at least until birth control became available[22].

Obviously being challenged with operation of equipment from a young age has an impact. Thus, the collective intellect of the human race increases. Use of a DVD, Ipod, Iphone, etc. is almost natural to today's youth and maybe not even comprehensible to a large percentage of people from a couple of generations past. The point is that evolution has taken place and continues to take place and even at an accelerated rate. The argument is made that these are weak measures of change in the human race. They may be called weak because they are subtle and it may be stated they are not correlated to changes in the DNA, brain size, loss of a tail, or genetics of the human species, but they are strong arguments due to the fact that they are measurable and deterministic. This is a short period of time for changes in a species but it did occur and that makes it a strong argument. Another problem that this brings up for IDers is that change does contradict ID. ID has to explain why are there any measurable changes. Intelligence of a level capable of creating life to the level of complexity, as we know it, would be a level capable of foreseeing the need for use of tools, appreciation of fine art, and immune systems that did not require constant attention. But these were changes that developed in humans over tens of thousands of generations, thus change from the original design was required. ID does not explain this change in humans.

The total accumulated knowledge of a grown adult from 10,000 years ago and today may be estimated. Anthropologist can estimate the knowledge base fairly close with study on behavior and environment. It is a large enough change that it may be estimated coarsely with some basic knowledge block, like vocabulary, tools used, education levels, number of associates, travel and geography knowledge, diet, eating, and health habits, and the list is quite lengthy. Religious leaders have dismissed this as a weak argument and say this does not support evolution. Regardless of labeling this as weak, it is a change that has occurred and a change that is not deniable. If people changed, then the original creation from ID required improvements. The human race changing with time contradicts the ID theory. The intelligence of this superpower to create life was very short sighted for a superpower.

The environment may cause changes that may be argued not relevant to the development of humans or that time was required to develop the environment we enjoy today. But environmental changes do induce change in the participants in that environment. Consider poultry raised in cages as compared to free running (aka free-range). Eggs from the same hen and rooster produce very similar adult poultry. If adult poultry raised from eggs of the same hen and rooster are separated after hatching and one confined to a cage for life as opposed to the other as free range, then they develop differently in the mental and physical characteristics.

Environment does drive change and especially in humans. As we change our own environment we create change in ourselves. Europe where dairy farming is prevalent, adults have developed a tolerance for dairy products and ability to digest dairy products. Africa and China tend to lag behind in dairy production and people consumed far less dairy products, therefore fewer people have a tolerance for lacteous. Malaria is prevalent in Africa and people are beginning to develop immune system capable of withstanding malaria and it is sad to say but through natural selection. As activity and labor-intensive work has decreased in just the past few generations, our cardiovascular systems are no longer adequate and people suffer what doctors call 'heart disease'. Some changes are so rapid that natural selection does not control future generations. People reproduce before heart disease hampers their ability to reproduce so the vulnerability to heart disease is passed to the next generation. Since humans are cognizant of this frailty and have the ability to change environment and habits, change will be created from intentional sources.

ID does not explain the diversity in the human race. If humans were one design created from a super-intelligent power, then why are there differences in bone structure, skin pigment, height, weight, and intellect? Some women are considered very beautiful and some very cute, while the majority women are not considered as attractive as the models selected for advertisements and TV news shows. Some men are predisposed or genetically capable of enlarging the muscular build with just exercise and some men require steroids while other men cannot build the Arnold Schwarzenegger look even with both steroids and exercise. Fingerprints, vein structure, face shape, and body proportions are different for all people. The differences in people prove that there was not one single design. Changes from the original humans prove that 'original design' was not perfect. There was not an original design of humans and there was not a perfect design of humans, which disproves the ID theory that a supernatural power created perfect human life, which did not require improvements.

ID theory is not well equipped to make biological comparisons, so IDers like to use analogies of mechanical devices. The comparison of a watch to a grandfather clock has been used to discredit the existence of the evolution of species. It is easy to comprehend that a crystal driven electronic watch did not evolve from a pendulum driven grandfather clock. The modern electronic crystal watch was designed with intelligence. IDers claim that because a watch did not evolve but was from intelligent design that the human race did not evolve but was created. Life is immensely more complex than watches and clocks. Humans designed mechanical and electrical devices and the knowledge base for different designs has changed since the initial building of devices to track time. While this is an interesting observation it is not analogous to the change of species in the trend of life. The argument could be that changes in design of time reporting devices supports theories of evolution. The intelligence changed from sundials, to water-flow, to sand-flow, to counting repetitious cycles and then the sophistication of repetition counting from mechanical to electrical pulses. Stating that a watch did not evolve from a grandfather clock is not evidence that life was created from ID! Presenting the obvious as a simple answer to dismiss the complex answer is palatable to religions because it endorses the religious system of belief and it is palatable to the intellectually limited constituency.

Humans did not co-exist with dinosaurs, and that is a fact. Campaigns have toured the country preaching that humans and dinosaurs inhabited the earth simultaneously. Children are taken on tours of museums and given the explanation that God created all creatures and the Bible says the earth is less than 10,000 years old so all creatures including dinosaurs lived in this time frame. This did not happen. Humans and dinosaurs are separated by over 60 million years. ID cannot explain the existence of dinosaur fossils. The existence of such creatures and then very rapidly becoming extinct does not suggest an inerrant creation. IDers freely admit that never finding human fossils with in the same sediment layers as prehistoric fossils is a problem but that it should be over looked. Science is not the practice of selecting facts to disregard that are not consistent with proposed theory.

ID uses the existence of our sight in two ways to indorse its doctrine by attempting to discredit evolution. One of the arguments is that there is no fossil trail of the eye development, the second is that the eye has to be complex and complete to function, so it could not be the product of evolution. The first fact is true that the eye does not fossilize but second fact is wrong and will be discussed later. Two facts do not make ID a science.

The eye is not fossilized due to the fact it is all proteins. It does not fossilize. The fact that there is no fossil trail of the development and evolution of the eye does not mean that sight was designed in for each species at its creation. No discovery of fossils does not mean that there is no evolution of sight. Tracing the evolution of sight is a difficult task and requires higher intellect than the ID theory creators are capable. The more primitive life forms such as worms that still exist do not have sight, as they were millions of years ago. Some insects that require sight for survival have sight. There is a direct correlation between the higher level of sophistication the life form and the better evolved the sight. Plants need to change their exposure to light depending on intensity. Sometimes they need to gather more and sometimes they need to avoid over exposure. This does not mean they have eyes, but they have developed a light sensitivity for improving the chances of survival. Light sensitivity is the most basic form of sight. Plants and animals split from the same life forms very early in the evolutionary process, shortly after multi-celled organisms developed so obviously their evolution was independent. Spiders and insects need to detect motion for survival and amazingly their eyes have evolved for this specialty. Birds and most animals need high resolution and better night vision but in some case do not require color distinction. It appears that with an "open mind" as the ID creators wish to instill, that a better case is made for evolution from the development of the eye. But this leads into the second claim of IDers.

Textbooks on ID claim that many components of the human body are too complex to function at reduced capability. In other words, if they are reduced or reversed in development they would not function. ID calls this presumption 'irreducible complexity'. This is incorrect. Some functions and organs appear to be irreducibly complex to casual observers but their perception is too shallow to be valid. Again, ID conjecture likes to select the eye as an example. ID claims that if any part of the eye is reduced it does not function therefore the eye had to be created, as it exists in today's animals. This cannot be examined without a search through the evolution of plants and animals. This one subject could easily fill a textbook, but a cursory review will demonstrate the errors of ID's assumption. People proposing and supporting ID cannot argue the biological and physiological qualities so they use analogies of mechanical devices. IDers like to use man-made devices as analogy to human evolution. ID uses the analogy of the camera to the eye to express their point of irreducibility. ID starts with a simple example to prove the point of irreducibility or a non-reducible design, the mousetrap. The mousetrap is used because it is easier to visualize functionally than a camera. Removal of any one part halts the function of the mousetrap. Then to increase in complexity, the clock and camera are presented as examples. If any part of the camera is removed it fails to function thus a camera cannot be developed by adding parts. First of all this is wrong, second it is a poor analogy and most important is that analogies do not prove facts. If a Single Lens Reflex camera has the shutter removed it does not work. This is true but not relevant to biological evolution. Another ID claim is the digital camera did not evolve from the SLR film camera. This is true it was designed. This is still not relevant to biological evolution. However, it could be argued that designs did evolve as designers learned from each progressive design from pinhole cameras to digital. But this is arguing the analogy, which is the wrong argument and not the subject of the evolution of the eye.

The eye did evolve and progress from very simple forms of life to humans and can be traced. Fossil trails are not required to follow this trail but it requires an understanding of evolution. This one subject on eyes could get deep enough to be its own course, but just a quick summary will reveal the fallacy of ID's assumption. Rather than examine the fossil trail, species still living today that developed millions of years earlier reveal the evolution. There are lower orders of life forms that 'see' light without the complexity of the human eye. Plants move to catch and avoid light with differing needs.

Agreed this is hardly sight but it is light detection and a plant is not an animal. Looking at animal progression from ancient species that still exist today but changed little from the original evolution shows the progression of sight. Many varieties of Phylum Platyhelminthes (soft body invertabrates flat worms) have light sensitive patches that detect light changes. These are not complete eyes but a start that it is probable that as new species evolved this was carried and improved as a start for eyes. Snails both fresh and salt water and terrestrial called limpets developed 'eye sockets' or slight indentations that provided two benefits. This cupped area for light sensitive cells add some directional and localize image detection. The Argonauta developed a deepened and more focused eye socket to improve on this directional and localize image detection. The next step that was quite large was the Annelidae that developed a transparent and protective cover to the 'eye'. Abalones or marine snails evolved fluid in the eye to improve focus. During the Jurassic period many land-dwelling dinosaurs return to the sea as amphibians. Fossil remains show the Stenopteryglus' eye sockets were for an eye about 10" in diameter. Today's Blue Whale has an eye around 3" in diameter. Stenopteryglus' was much smaller than a Blue Whale of today but yet had a much larger eye. This does not prove beyond a doubt that its eye was less sophisticated but is very strong evidence. Squids have existed for millions of years with little change. Squids change colors to display disposition and communication. They use this as warnings of territorial invasion or for attracting mates. Their eyes are far less sophisticated than mammals but very useful for determining color for communication. As fish, reptiles, and mammals evolved the eye was not to human standards. Some species developed eyes especially suited to their survival, such as seeing in ultraviolet spectrum for tracing the suns movement or seeing in inferred for heat detection. But clearly the use of the eye has assisted in the survival and changed during the last 250 million years. Surviving species define the evolution and the eye provides more support for evolution than it does for ID. This discussion could continue but it clearly shows the eye is not an example of a non-reducible design. The immune system is another example raised by IDers. Again, this does not leave a fossilized trail. The immune system will show the same results if time is taken to trace the history of development. The same case can be made through living organisms and facts about the human body. Primates, human ancestors, had the ability of making vitamin C and lost it through time as they ate fruits that provided vitamin C. Evolution is not the point here that is for biological science, the point here is that non-reducible design is not true and does not support ID.

Many IDers do admit evolution has occurred and does occur on a micro-scale. IDers are willing to concede that some species existence is only explained by evolution. Species that live on islands have evolved slightly different traits than those on continents. They are most similar to species on the closest mainland but slightly different. For these cases IDers concede that evolution is the only plausible explanation. But IDers try to set limits that this does not relate to human beings. Setting a limit on evolution has two flaws. First is that the limits are set arbitrarily to not interfere with religion rather than from any physical law. Second is that setting an arbitrary limit implies their intelligent design creator was a prankster. Why would he have lower species evolve and not humans? Admitting these island species as evolved creates a minimum time for the evolution that far exceeds their 10,000-year limit they place on earth's existence.

IDers have tried a second approach to eliminate the microevolution as they call it. Again, IDers like to use mechanical devices as an analogy to slight variations in species. The examples are that designers and engineers may use several variations a design for the same function. Automobiles, clocks and watches, TV sets, radios, and eyeglasses can all appear different and perform the same function. This is trying to use the analogy of an engineer selecting variations of the same design for slightly differing circumstances and applications. IDers use this argument to state the ID created variations in the different species for different locations around the world. The creation of variations of species in differing geological areas again shows the creator as a prankster. Why did the creator only place special species where they were most likely to survive? Also why would the prankster not create land animals on inlands they could not reach by their own means? Science explains the species of insects and animals found on islands and their nearest continent or the continent that air and water currents would have provided a means of relocation. This does not prove the IDers wrong because it is impossible to prove something did not happen. However, stating that ID provides an answer by stating the creator chose options and slight variations is an extreme reach only supported by faith. No evidence is available to support the idea that a creator chose to make several variations of fruit flies in Hawaii that are slightly different from other locations.

Proponents of ID claim it has to be true and answers the question of how we became the life form that we are, because evolution has missing links. In other words, there are steps in the evolutionary process that are not proven with artifacts and fossils therefore ID is true. The argument that there are missing links continues to diminish as new discoveries are made. Fossils have been found of fish that had fingers and elbows on its front fins that could have allowed it to crawl along the ocean floor. The same fish had a bone structure the allowed its head to turn side to side like many animals and unlike other fish. This is different from reptiles and mammals that returned to the seas, such as dolphins and whales. This one finding does not fill in the final missing link but it eliminates many of the missing steps. Every step in the evolutionary process may not be found. People with alternative agendas to science will most likely always be able to find a 'missing link'. That does not prove evolution to be false. Also, having missing steps in the evolutionary process does not support an alternative explanation for our existence. ID has to have scientific reasoning to be viable. ID just discrediting other sciences is not reasoning. Denying that evolution exists is being dishonest. Replacing evolution with ID is also dishonest.

The largest flaw in ID is the design itself. We are not perfect. Many species are not perfect. A creator with the intelligence to create all forms of life would have the intelligence to have eliminated some of the problems that plague us. All species have some flaws. Was this again a hoax by a master prankster with a sense of humor that exceeds all comedians? More species have gone extinct than exist. Did the great creator make a mistake on these species? Many more species will go extinct due to man made changes in the environment and human expansion. Survival of the fittest has been dependent on food availability, protection or evasion of predators, and reproduction exceeding demise. The more recent constraint has been added of coexistence with humans. The creator did not foresee such an event in creating animals. Humans show frailties that have no purpose. Human cancer rates increase with exposure to carcinogens. There is no benefit to living by developing cancer from asbestos, cigarette smoke, benzene, etc. The human eye develops cataracts with time and exposure to UV light. Nearly all humans need corrective lens for vision improvement. The eye alone provides many examples of imperfect design. Our immune system that IDers wish to point out as irreducible design also shows many frailties. We are susceptible to many diseases from bacteria and viruses. Some illnesses for which the immune system cannot defend against or protect our bodies result in death. Human immune systems have provided the opportunity for survival. It is not perfect and could not be considered designed by the ultimate intelligence, but it fits well into the survival of the fittest pattern. The appendix is a ruminant of a long-term digestive track not required as our diets changed. The change in diet was from hunter-gatherers starting hundreds of thousands of years ago when the foods provide substance for living without long-term digestive tracks. The appendix is now an organ to be inflamed and infected causing illness. The appendix does not provide creditability to the ID explanation. Human being and primates require a food source of vitamin A and C. We no longer have the ability to produce our requirements of vitamin A and C. Many mammals can produce their own needs of these but humans have lost the ability. Humans are susceptible to depression and emotional distress. It seems a more intelligent design would allow people to control their emotions and certainly not commit suicide from mental anguish. Diabetes, heart disease, and major illnesses that shorten lives show the susceptibility of humans and do not add evidence for ID.

The final conclusion drawn from analyzing intelligent design is that life does not represent the level of intelligence required to create it and there is no factual evidence to support it. ID is a strategy to displace sciences that are not consistent with religious beliefs. The table that follows is to summarize the tenants of Intelligent Design and give the facts relating to each facet:

Comparison of ID tenants to the truth	
ID is science.	Scientific analysis is strictly factual. Biological sciences require learning and understanding life from facts, not beliefs or faith. ID is politically motivated for an agenda other than science.
ID is simple and straightforward.	Science is extremely complex and requires millions of scientists and many generations to collect and analyze data and develop conclusions and theories.
Many Christian leaders purport that the earth is less than 10,000 years old and conforms to ID theory.	The earth is 4.5 to 4.6 billion years old, and the universe exceeds 13.5 billion years.
Many Christian leaders purport that evolution is a "theory" and not proven, and thus should not be considered as true.	Evolution is fact and has millions of proofs. IDers try to use theory as a suggestion of whimsical, not the scientific usage as in the theory of gravity.
Many Christian leaders purport that ID should be taught in place of evolution.	ID is a religious ploy not a science and should be carefully scrutinized and certainly not displace science.
Many Christian leaders purport that ID should be taught in parallel with evolution and students should keep an open mind to choose the correct answer.	Evolution and ID are mutually exclusive and cannot be taught that both are true. Students should be taught true facts and not beliefs or faiths.
IDers have used emotional pleas and illogic to gain entry into public schools, like a fair and balanced approach.	Schools should only teach factual information. The Columbine shooting was not a result of teaching evolution.
ID is not from the Bible but maybe a wedge strategy to gain entry into public education	ID is not listed as such in the Bible but is carefully constructed to parallel the Biblical creation and allow semantically the substitution of a supernatural being in place of God.

Comparison of ID tenants to the truth	
ID was created in the 1980's, as a result of efforts to force creationism into public schools that were not successful.	Recently being developed does not make ID true. Failure of evangelists to get creationism to replace science has spawned ID to reduce contradictions of religions by science.
ID is simple and easy to understand and evolution is extremely complicated.	This is true but only relies on peoples' laziness. Being simple does not add creditability to ID.
ID claims to be a rational, defensible, well-argued Judeo-Christian worldview.	Religions are millenniums old but that does not relate to correctness or being true.
People co-existed with dinosaurs.	Human beings and dinosaurs are separated by more than 63 million years. "Jurassic Park" and "The Flintstones" are fictional.
Humans are consistent and non-changing.	No, people are constantly changing, but changes require a very precise statistical study to determine changes due to the short time frame of the sample.
Our grandparents did not resemble chimpanzees, so people did not evolve from primates.	People did evolve from primates and it has been proven with fossils and DNA similarity.
ID recognizes mini-evolution in insects, small animals, bacteria and viruses, but claim humans did not evolve but were created, as they exist today.	Mini-Evolution such as viruses and bacteria strains changing have been documented and cannot be denied so ID acknowledges these, but ID tries to exempt humans' changes as conflicting with the Bible. ID's claim does not change the evidence of human evolution and does not explain the limitation of evolution for human beings.
ID states species were designed, as they exist.	ID misinterprets the use and meaning of survival of the fittest, natural selection and artificial selection.

Comparison of ID tenants to the truth	
Teaching evolution resulted in the Columbine school shooting and many other atrocities.	No. If statistics are used to analyze all school shooting there is only one common factor, all shooters used guns. There is no reference to any factor relating commonality to any academic requirements.
ID's central point is that humans did not evolve from lower intellect animals.	This is wrong and all evidence shows this to be wrong.
ID claims there are missing links in evolution	Correct, but these are diminishing with new discoveries yearly. Missing data does not make evolution wrong or ID correct.
Parts of the human body are "irreducibly complex" and had to be design as they are to function. There would be no functional reduced part that would evolve to meet the present function.	This is wrong. What ID calls "irreducibly complex" can be shown to be reduced even in the species that have existed for millions of years but maintain their original function and many are alive today.
Not enough evidence of evolution is found in fossil remains to prove humans evolved from apes.	There is substantial fossil evidence to show humans evolved from apes and supporting data like sediment layers and DNA to verify fossil findings.
ID uses mechanical analogies to explain designs of humans from intelligence design (God).	Analogies are not proof. Analogies are to direct thinking and generate rhetoric to avoid analysis.

Comparison of ID tenants to the truth	
ID claims humans to be unique and completely separate from all other animals (also a way of saying humans have a soul and other animals do not).	This is an attempt to drive biological science beyond the limits of public school. The answer is yes humans are different. Our DNA is not a perfect match to any other species. The sum of all human traits is not found in any other animal. Many traits for socializing, hunting, reproduction, and even some emotions are found in primates and other animals. While humans are unique with all collective traits, the traits are not exclusive. The intent of the statement is that humans share nothing with lower animals and therefore could not have evolved from them is not true.
Humans are the ultimate design.	Not true. We have many flaws and we continue to evolve. If humans change the world to eliminate the human race, other species will evolve and it cannot be said that the new species will not exceed human capabilities. We are not the ultimate evolutionary product.
Fossils of dinosaurs and humans were placed in different sediment layer to fool scientists.	No, the ultimate hoaxers or prankster would have placed human remains in the same sediment layers as prehistoric fossils to make the joke or riddle more complicated. This is not the case, dinosaur and human fossils are in different layers because of the time difference in their existence.

Questions over Chapter 16

1. ID is simple and evolution is very complex
   a. proves ID is correct and evolution is wrong.
   b. is a true statement but proves nothing about the truth of either.
   c. is the best reason to teach ID.
   d. proves evolution is correct and ID is wrong.
2. The different species with differing levels of sophistication and capabilities
   a. prove all life was designed for its own purpose.
   b. disproves evolution.
   c. has no relevance to intelligent design.
   d. is very difficult to explain and requires more investigation than this course can cover.
3. Our similarity to past generations of humans is
   a. exact since there has been no change.
   b. very difficult to measure because tests were not performed on earlier generations.
   c. obvious and makes ID correct.
   d. obvious but the differences are insignificant and should be ignored by science since they are small.
4. Intelligent design cannot have been the source of our world because the intelligence to create life
   a. is still not known.
   b. does not show in our vulnerabilities to disease.
   c. because the human race has changed measurably in the last 10,000 years.
   d. all the above.
5. Comparing a watch to grandfather clock shows
   a. nothing because it is an analogy not evidence.
   b. the electronic watch was designed from the grandfather clock.
   c. that humans could not evolved from monkeys.
   d. that humans could have evolved from monkeys.
6. Missing links in the evolutionary chain proof
   a. intelligent design is the answer.
   b. that science has not discovered every connection.
   c. that religion is right and science is wrong.

    d. religious leaders are creating stories to discredit science.

7. ID explains dinosaurs
    a. by stating both humans and dinosaurs coexisted simultaneously.
    b. by stating God placed fossils in sediment layers.
    c. by taking children on tours of museums and explaining all creatures were designed.
    d. all the above are answers but not necessarily true.
8. The eye and especially in human beings shows
    a. ID has to be correct since we can see.
    b. ID has the only explanation for sight.
    c. evolution can explain life even without fossil trails.
    d. both ID and evolution need to be considered.
9. Irreducible complexity means
    a. a function or organ will fail to function if it is reduced in complexity.
    b. eyes are like cameras.
    c. humans could not have evolved.
    d. all the above.
10. ID explains differences in species
    a. as errors in science studies.
    b. the creator selected different options.
    c. as superficial and insignificant.
    d. a trick by evolutionists.
11. ID explains human imperfections
    a. because they are God's will and we are not capable of understanding.
    b. as pranks by a master prankster.
    c. as insignificant.
    d. is wrong in that ID does not explain human imperfections.

12. Understanding the science of how the human race came to exist is important
    a. to accept all biological sciences.
    b. to understand where we came from to provide knowledge of where we are headed.
    c. to defend science from religious attacks.
    d. all the above.

Probability of statements in Chapter 16

Instructions: Assign a number to the plausibility of each statement.

(1) For a total contradiction to a statement.
(2) For an implausible statement.
(3) For a possible and maybe even practical statement.
(4) For a plausible statement.
(5) For a confirmed or verifiable statement.

1. The Bible does not mention intelligent design. Intelligent design is a recent contrived theory to displace science. What is the probability that intelligent design is fact? (1 – 5)?
2. Intelligent design claims to be the answer, because it claims we were made perfect and have not changed. What is the probability that that we have not changed? (1 – 5)?
3. What is the probability that intelligent design answers how humans were created? (1 – 5)?
4. What is the probability that intelligent design is true because the theory of evolution has missing links? (1 – 5)?
5. What is the probability that intelligent design is true because TV evangelist say it is? (1 – 5)?
6. What is the probability that knowing that we were designed by an intelligent being makes us safe from self-destruction? (1 – 5)?
7. What is the probability that people who are giving warnings about the human race destroying the environment are alarmist trying to make money from scaring the public? (1 – 5)?
8. What is the probability that understanding the beginning of humanity is not important? (1 – 5)?

Glossary of terms and names:

Age of Enlightenment – Middle to late 17th century, started approximately 1650.

Age of the Dark Ages – The start of the 6th century to the 11th century, or from approximately 500CE to 1000CE.

Age of the Middle Ages – The age between the Dark ages and the Age of Enlightenment of approximately from the 11th century to the middle of the 17th century, or approximately from 1000CE to 1650 CE.

Akkad, Akkadian – Akkad was the city and capitol on the Akkadina empire. The city was on the west side of the Euphrates and southwest of present-day Bagdad, Iraq. The empire rose to power around 2400BCE and declined after approximately 200 years.

Apocrypha is text documents and book of spiritual writings left out of the Canonization of the Bible. Many documents were believed to exist in libraries that were burned and destroyed such as the library of Alexandria.

Arabia – The peninsula that is now Kuwait and Saudi Arabia and part of Iraq and Syria. The waters surrounding the Arabian Peninsula are best known Red Sea, Persian Gulf and the Gulf of Aqaba. The Arabian Sea (part of the Indian Ocean) is on the south and on the northeast the Gulf of Oman and the Strait of Hormuz.

Assyria, – The empire that ruled the northern half of Mesopotamia (with Babylonian rule of the southern half). The capitol city was Assur. The empire built up slow from about 2000 BCE and started a decline in power around 900 BCE and was virtually nonexistent by 600 BCE. Hammurabi was an early ruler of Assyria and complied the first known written laws.

Assyrian Flood Myth is from the Tablet 11, from the Atrahasis Epic into the Gilgamesh Epic was written around 2000 to 2100BCE. Tablet 11 contains the flood story.

Atrahasis Epic contains Tablet 11of the Assyrian Flood Myth of the Gilgamesh Epic.

BCE – the era before current time keeping was started at zero, or 2010 years before this book was published. Years are given before the restart at 0, which religions like to correspond to the commonly accepted birth of Christ. The time 2000 BCE would correspond to the Christian marking of 2000BC.

Babylonia, Babylonian The empire that ruled the southern half of Mesopotamia (with Assyrian rule of the northern half). The empire centered around Kish and Babylon. The first record of Babylon was from 2300 BCE and was the tablet of Sargon of Akkad (ruler).

Canaan refers to the region between the Jordan River and the Mediterranean Sea and from northeast Egypt to Lebanon. Canaan is located on the Southern portion of the East end of the Mediterranean Sea. Today this area is Israel, Jordan, Lebanon, Palestinian Territories, and parts of Egypt and Syria.

Charles Darwin was an English Naturalist that lived from 1809 to 1882. He studied medicine at University of Edinburgh in Scotland. He then studied theology at Cambridge. He is credited with initiating the ideas of differing species evolving from common ancestry through natural selection. Charles Darwin authored ***The Descent of Man, and Selection in Relation to Sex*** and ***On the Origin of Species.***

Cherubims – Spiritual winged guardians with lower status than angels. As with angels, no evidence has been found of actual existence. Most common is winged body a lion, bull, or horse although it may be bipedal (as human) and possess the head of an eagle, lion, ram, bull, or many times human. Many different drawings have portrayed the believed representation of cherubim from the many Biblical references. Cherubims are as well defined as angels and like angels have never been documented.

Cubit is the ancient measurement and commonly thought of as 1½ feet, 18 inches, or one-half yard.

Cush, Cushites Biblical reference is often confused with historical placement. Cush in the biblical reference applies to the King Cush and the land he ruled. Cush may have been confused with or legends barrowed from Kush, or the Kushites. Kush was a city that rose as a small empire before the Egyptian empire became dominant. Kush was a city in the Nile River valley in what is now Northern Sudan. Cush is intended from Biblical indications in the northern Mesopotamian plain and on a tributary to the Euphrates, while Kush is south of Egypt near the Nile.

Dead Sea Scrolls the ancient documents discovered in 1947 in caves in the Qumran desert at the North West bank of the Dead Sea. Discovery did not complete until 1979 and reconstruction and translating still continues even though most is complete. Approximately 1000 documents were discovered and at least a small part of every book of the Old Testament was found. No part of any book of the New Testament was found although some clergy have claimed some fragments to contain a few letters that could have been used in the New Testament.

DNA is the acronym for DeoxyriboNucleic Acid. The DNA molecules are nucleic acid is a long polymer of basic nucleotides that contain the genes defining the life form. This is a very brief explanation and further research by Internet searches is suggested.

Egypt, Egyptian – The land of present-day Egypt and a civilization that started around 3200 BCE. Civilization was centered on the Nile River.

Ethiopia, Ethiopian – Country east of Sudan and South of Egypt in east Africa. It is one of the old civilizations and may have influenced Biblical writings or it is possible that translation error labeled people as Ethiopian that were named for the ruler of Cush.

Euphrates River one of the two rivers developing civilization in Mesopotamia. Euphrates still flows through central Iraq and is west of the Tigris River. North of Basra, Iraq the Euphrates merges with the Tigris to form the Shatt al-Arab to form a short river that empties into the Persian Gulf.

Firmament is commonly interpreted as the heavens or dome above earth. The Firmament was considered a rigid structure that held the sun, moon, and stars. It was considered by the Jewish religion as separating heaven and earth and provided a domain for God.

Gilgamesh Epic see the Assyrian Flood Myth and Atrahasis Epic above.

Gnostic Gospels were found all at the same time by two farmers in the Nag Hammadi area of Egypt in December of 1945. About one third of the find was burned for fuel. The remaining scrolls have been preserved and translated. The Gnostic Gospels were written as the teachings of Jesus from about 100CE to 200CE or perhaps as late as 300CE.

Gutenberg German Bible was the German translation from the Vulgate or Latin Bible and printed by Johannes Gutenberg on the moveable press printer in 1452 and 1453.

Hammurabi's law – Hammurabi was the sixth king of Babylon and was a First Dynasty king of Babylon. He extended control to cover most of the Mesopotamian plain. His successors were unable to maintain control over this large are. Hammurabi recorded the first written or at least the oldest known written laws or codes. The stone tablet containing these laws was discovered in 1901, was over six feet in length and stored in the Louvre Museum in Paris.

Isaac Newton - Title: Sir Isaac Newton was and English writer, physicist, mathematician, astronomer, natural philosopher, alchemist and theologian. Sir Isaac Newton in 1687wrote the *Philosophiæ Naturalis Principia Mathematica*, which helped start the science and mathematics revolution and age of enlightenment.

Mesopotamian is the lands covering Iraq, western Iran, Northeastern Syria, and Southeastern Turkey.

Nile is a river in the eastern 1/3 of Egypt with tributaries that form in Tanzania and Ethiopia that combine in Sudan and flow basically North to the Mediterranean Sea.

Ossuary of James, brother of Jesus is a hoax that was created from an actual ancient burial box from approximately 2000 years old. The "brother of Jesus" was added to the original engraving and has been verified as fraudulent and analyses are available on Internet searches.

Papyrus is a verity of materials made from fibers from the papyrus plant in Egypt. Long strips were woven and compressed together to form material and one use was as paper.

Parchment is writing media (paper) made from calfskin, sheepskin or goatskin. It easily formed scrolls for storing writing such as the books canonized into the Bible.

Pentateuch or Torah was the first five books of the Old Testament and commonly attributed to Mosses as the author.

7Q5 fragment from the Dead Sea Scrolls is a fragment that is claimed by a Jesuit Priest, Jose O'Callaghan to be part of the New Testament from the Dead Sea Scrolls. His claim is based on a fragment that may possibly contain 13 letters and only 8 are certain, and only ONE complete word, which translate as a conjunction like "and". NO other fragments of the millions recovered are thought to contain any portion of the New Testament. Fragment 7Q5 is obviously a fraudulent analysis from wishful desire. The designation refers to Qumran cave7 group Q5 for sorting fragments' locations. This and many results of the Dead Sea Scroll fragments results are available on Internet searches.

Qumran refers to the caves and community living near the Northwest corner of the Dead Sea from around 150BCE to around 100CE (the community may have been lost as early as 70CE after the fall of Jerusalem and at the time of the destruction of the temple in Jerusalem). Qumran caves were discovery location of the Dead Sea Scrolls. The community was believed to be the Essences sect of the Hebrew faith. A great deal more information is available from the Internet and books on the Dead Sea Scrolls and Qumran.

Renè Descartes French philosopher considered the Father of Modern Philosophy. He was a writer, scientist and mathematician and a key figure in the enlightenment or scientific revolution.

Septuagint is the Greek Bible and named "**LXX**". The Greek version was translated between the 3BCE to 1BCE in Alexandria from the Ancient Hebrew bible.

Seven deadly sins are also known as the Cardinal Sins or Capitol Vices. These were not in the Bible but created by the Catholic Church. These started to evolve from the 4th century from monk Evagrius Ponticus' list on evil thoughts believed to contrast the virtues with of "Galatians" in the *New Testament.* They were refined by Pope Gregory in the 6th century as part of church code for Europe.

Shroud of Turin was claimed to the burial cloth covering Jesus after his execution. The cloth was actually made around 1400 CE and even denounces by the Pope at that time to be a hoax. A plethora of science tests have proven it to be a hoax and available on Internet searches.

Spiritual facts are those contradictions in religion to other religious facts or to reality. Religious leaders justify the contradictions by spiritual facts with many phases, such as only God knows, God will reveal this when we get to heaven, human beings cannot understand, etc.

Sumer, Sumerian – The earliest known civilization. Sumer was in southern Mesopotamia and began civilization around 5500 BCE.

Tanakh was the text or Bible assembled around 2600 years ago and used as the religious text by the Jewish Religion or Judaism. There have been some fragments found of some of the books on pottery as old as 3000 years ago dating back to King David's reign. This contained the Torah or Pentateuch and Nevi'im ("Prophets") and Ketuvim ("Writings") and thus the combined name Ta- Na – Kh. The Tanakh was 26 books that makes most of the Old Testament.

The King James Version of The Holy Bible: The translation of the Bible authorized by the Church of England commanded by King James the 8th, published in 1611. This was to make available the English translation from Masoretic Hebrew text, the Greek Septuagint (LXX), and the Latin Vulgate and compared to the Gutenberg (German Bible of 1455).

Tigris River one of the two rivers developing civilization in Mesopotamia. Euphrates still flows through central Iraq and is east of the Euphrates River. North of Basra, Iraq the Tigris merges with the Euphrates to form the Shatt al-Arab to form a short river that empties into the Persian Gulf.

Torah much the same as first five books of the Tanakh or Pentateuch was the first five books of the Old Testament and commonly attributed to Mosses as the author.

Vulgate is the early Latin Bible commissioned by Pope Damasus I, in 382 and is primarily the work of Jerome. The Vulgate was translated from the Hebrew Tanakh for the Old Testament, which contained 46 books. The 27 books forming the New Testament is much more difficult to determine the sources that Jerome used.

Notes

Notes and explanations denoted in text.

---

[1] Braille is the printing method for reading through physical touch invented in 1821 Frenchman Louis Braille.
[2] Original writings that are included in the Old Testament part of the King James version of the Bible have been dated to the 10th century BCE. The text was found on a scribe portion of pottery similar but not exact copy of the text in Isaiah 1:17, Psalms 72:3, and Exodus 23:3. These were not the same text as in the Torah.
[3] Inconclusiveness is meant to indicate doubt has been tested to the extreme but not proven.
[4] Note these are not unique areas and have some common areas but are enclosed from North Africa to India to southeastern Europe.
[5] The Bible is the only reference for this text. This is information to guide the student in search for the source of the story of the trees of life and knowledge. This is given as information for Internet searches and not a reference for contradicting the Bible passage.
[6] Please excuse my jumping ahead in the Bible story to where God named Adam, but I felt for consistency in analysis of the story it could be mentioned here.
[7] The moral principle that men are superior and dominant over women is either the humans writing the Bible and injecting their beliefs *or* it is the inerrant Word of God. Rush Limbaugh has been the self-appointed voice of moral decision on gender roles. Students can search the internet for R. Limbaugh's number of failed marriages, limited educations, drug abuse and scandals, and in general his lack of moral content. R. Limbaugh epitomizes and demonstrates clearly the fallacy of the concept of gender superiority. God's Word providing parallel gender roles with that of R. Limbaugh would not be consistent with the power and intelligence associated with God.
[8] Since the only resource for this textbook is the King James Version of the Holy Bible, Egyptian Assyrian and Babylonian mythology and religion are not covered and left for inquisitive students to research.
[9] Assuming in this primitive world that they stay together for survival after being driven from the garden.
[10] Noah is of the tenth generation on earth, but Adam and Eve only had Seth toward populating the world and Noah's or the tenth generation has not been multiplied in as having started the next generation.
[11] Religious leaders oppose any science or thought that promotes skeptical inquiry.
[12] Most commonly accepted measurement is 18 inches or 0.457 meters for a cubit.
[13] The Great Pyramids in Egypt are 420-foot tall by 660-foot base or in original measurements 280 cubits in height by 4x440 cubits. This geometry

would achieve greater height than geometry of a tower.

[14] Let's refer to Intelligent Design as ID for ease of reading, writing and conciseness and use ID to indicate the theory of creation by an intelligent superpower.

[15] Please allow the abbreviation and grouping of all proponents supporting teaching ID as a science as IDers.

[16] The reason exposing the truth about ID or any religious subject conflicts with religion is that religious leaders contend that there are spiritual truths that cannot be verified by humans and only explained by God.

[17] Tom Delay was in the US House of Representatives as a Republican from Texas from 1984 to 2006 and was House Majority Leader from 2003 to 2005.

[18] Religious tolerance in this sense refers to the allowing views that are not those of Christian Evangelist and conservatives.

[19] During the Middle Ages and later during the Enlightenment period philosophers and scientist were punished, tortured, and even executed for claiming space was a vacuum or even that a vacuum could exist.

[20] DNA testing yield the similarities and differences in genes that govern all aspects of each individual animal's characteristics, which determines species exactly and even the difference within species.

[21] Cave drawing date back 27,000 years and science has been actively attempting to understand the origins of life for less than 200 years, which defines the term of recent times.

[22] Birth control is another huge change in the human race, showing it is evolving by creating by its own selective reproduction. The effects will be seen in time and difficult to predict.

www.ingramcontent.com/pod-product-compliance
Lightning Source LLC
LaVergne TN
LVHW010106170826
845678LV00012B/2266

9798375650463